Protein sequencing

a practical approach

TITLES PUBLISHED IN

THE PRACTICAL APPROACH SERIES

Series editors:

Dr D Rickwood
Department of Biology, University of Essex
Wivenhoe Park, Colchester, Essex CO4 3SQ, UK

Dr B D Hames
Department of Biochemistry, University of Leeds
Leeds LS2 9JT, UK

Affinity chromatography
Animal cell culture
Antibodies I & II
Biochemical toxicology
Biological membranes
Carbohydrate analysis
Cell growth and division
Centrifugation (2nd Edition)
DNA cloning I, II & III
Drosophila
Electron microscopy in molecular biology
Gel electrophoresis of nucleic acids
Gel electrophoresis of proteins
Genome analysis
HPLC of small molecules
HPLC of macromolecules
Human cytogenetics
Human genetic diseases
Immobilised cells and enzymes
Iodinated density gradient media
Light microscopy in biology
Lymphocytes
Lymphokines and interferons
Mammalian development
Microcomputers in biology
Microcomputers in physiology
Mitochondria
Mutagenicity testing
Neurochemistry
Nucleic acid and protein sequence analysis
Nucleic acid hybridisation
Oligonucleotide synthesis
Photosynthesis: energy transduction
Plant cell culture
Plant molecular biology
Plasmids
Prostaglandins and related substances
Protein function
Protein sequencing
Protein structure
Spectrophotometry and spectrofluorimetry
Steroid hormones
Teratocarcinomas and embryonic stem cells
Transcription and translation
Virology
Yeast

Protein sequencing

a practical approach

Edited by

J B C Findlay

Department of Biochemistry, University of Leeds, Leeds LS2 9JT, UK

M J Geisow

Delta Biotechnology, Castle Court, Castle Boulevard, Nottingham NG7 1FD, UK

IRL Press
Eynsham
Oxford
England

First published 1989

British Library Cataloguing in Publication Data

Protein sequencing.
1. Peptides & proteins. Sequences. Chemical analysis
I. Findlay, J.B.C. II. Geisow, Michael J. III. Series
547.7′5046

Library of Congress Cataloging-in-Publication Data

Findlay, J. B. C. (John B. C.)
Protein sequencing.

Includes bibliographies and index.
1. Amino acid sequence—Congresses. I. Geisow, Michael J. II. Title.
QP551.F65 1988 547.7′5 88-32025
ISBN 0 19 963012 7 (hardbound)
ISBN 0 19 963013 5 (softbound)

Previously announced as:
ISBN 1 85221 090 7 (hardbound)
ISBN 1 85221 091 5 (softbound)

Typeset and printed by Information Printing Ltd, Oxford, England.

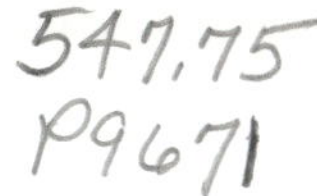

Preface

The ability to determine the primary structure of a polypeptide was a major advance in biochemical sciences and sequencing work since then has had an important impact on our understanding of protein structure and function. At present, protein sequence analysis occupies a position at the forefront of modern molecular biology. Partial sequence determinations using microgram amounts of protein has enabled many genes with low levels of transcription to be cloned and their complete primary structures to be determined. This approach is also the only means available for determining the sites of post-translational modification.

This book is aimed at those wishing to isolate proteins or peptides and to carry out subsequent sequence analysis. It is written for both professional and inexperienced research workers and for those with and without access to high technology instrumentation.

At the start, much space is given over to the protein and peptide purification with particular emphasis on the important techniques of microscale HPLC and electrophoretic/electroelution techniques. The next chapter details a wide range of methods available for amino acid analysis, cleavage and detection of polypeptides and protein modification (also included as an appendix). The protocols described have been carefully selected for their general usefulness.

The sensitivity and efficacy of sequencing procedures have improved dramatically over recent years due to the development of new reagents, new protocols, new strategies and new instrumentation. The most effective modern techniques are detailed in sections dealing with manual, gas-, liquid- and solid-phase sequencing and with the rather different approach utilizing mass spectrometry.

The information for the folding and three-dimensional structure of a protein is contained in its amino acid sequence and considerable attention is being paid to the difficult task of deciphering this code. The final chapter of the book illustrates the principle and approaches to structure prediction and provides methods which can be used without recourse to sophisticated computer programs and hardware.

As in all the 'Practical Approach' volumes, as much detail as possible has been given together with critical references which illustrate and extend the methods described. Sequence analysis is in a very active growth phase and the competition, particularly in the commercial sector, to improve chemical approaches, separation systems and sequence instrumentation is very keen. We would be very interested to hear, therefore, of advances or corrections which would improve subsequent editions of this text.

J.B.C.Findlay
M.J.Geisow

Contributors

A.Aitken
National Institute of Medical Research, The Ridgeway, Mill Hill, London NW7 1AA, UK

K.Biemann
Department of Chemistry, Massachusetts Institute of Technology, Cambridge, MA 02137, USA

J.B.C.Findlay
Department of Biochemistry, University of Leeds, Leeds LS2 9JT, UK

M.J.Geisow
Delta Biotechnology, Castle Court, Castle Boulevard, Nottingham NG7 1FD, UK

C.Holmes
Department of Biochemistry, University of Dundee, Dundee DD1 4HN, UK

J.N.Keen
Department of Biochemistry, University of Leeds, Leeds LS2 9JT, UK

D.J.C.Pappin
Department of Biochemistry, University of Leeds, Leeds LS2 9JT, UK

W.R.Taylor
National Institute for Medical Research, The Ridgeway, Mill Hill, London NW7 1AA, UK

J.M.Thornton
Department of Crystallography, Birkbeck College, Malet Street, London WC1E 7HX, UK

K.J.Wilson
Applied Biosystems, 850 Lincoln Center Drive, Foster City, CA 94404, USA

A.Yarwood
Department of Botany, Science Laboratories, University of Durham, South Road, Durham, DH1 3LE, UK

P.M.Yuan
Applied Biosystems, 850 Lincoln Center Drive, Foster City, CA 94404, USA

Contents

Abbreviations

ACTH	adrenocorticotropin
AEAPG	*N*-(2-aminoethyl)-3-aminopropyl-glass
AP	3-aminopropyltriethoxysilane
APG	aminopropyl-glass
ATZ	anilinothiazolinone
CAP	catabolite activator protein
CID	collision-induced decomposition
CPG	controlled pore glass
DABITC	4,*N*;*N*-dimethylaminoazobenzene-4′-isothiocyanate
DABTH	dimethylaminoazobenzene thiohydantoin
dansyl	see DNS
DITC	*p*-phenylene diisothiocyanate
DMA	dimethylamine
DMF	dimethylformamide
DMPTU	*N*-dimethyl-*N*′-phenylthiourea
DNS	5-dimethylaminonaphthalene-1-sulphonyl
DPTU	*N*,*N*′-diphenylthiourea
DPU	*N*,*N*-diphenylurea
DTNB	5,5′-dithionitrobenzoic acid
DTT	dithiothreitol
EDC	1-ethyl-3(3-dimethylaminopropyl)-carbodiimide
EDTA	ethylenediamine tetraacetic acid
EGF	epidermal growth factor
FAB	fast atom bombardment
FMOC	*N*-(9-fluorenylmethoxycarbonyl)
GLC	gas–liquid chromatography
GOR	Garnier–Osguthorpe–Robson
HPLC	high-performance liquid chromatography
MHC	major histocompatibility complex
MS	mass spectrometry
NBS	*N*-bromosuccinimide
NEM	*N*-ethylmorpholine
NTCB	2-nitro-5-thiocyanobenzoic acid
OPA	*o*-phthalaldehyde
PAGE	polyacrylamide gel electrophoresis
PD	plasma desorption
PIR	protein identification resource
PITC	phenylisothiocyanate
PMSF	phenylmethylsulphonyl fluoride
PSQ	protein sequence query
PTC	phenylthiocarbamyl
PTH	phenylthiohydantoin
PVDF	polyvinylidene difluoride
QA	*N*-trimethoxysilyl-propyl-*N*,*N*,*N*-trimethylammonium chloride
RP	reverse-phase
SDS	sodium dodecyl sulphate
TCA	trichloroacetic acid

TEA	triethylamine
TETA	triethylenetetramine
TFA	trifluoroacetic acid
THF	tetrahydrofuran
TLC	thin-layer chromatography
TMA	trimethylamine
TNBS	2,4,6-trinitrobenzenesulphonate

CHAPTER 1

Protein and peptide purification

KENNETH J.WILSON and PAU M.YUAN

1. INTRODUCTION

The role of the protein chemist in the characterization of either natural or recombinant peptides or proteins has steadily gained in importance. Whereas some reviewers at the end of the 1970s questioned the need of continuing efforts in structural determination, one now suspects that a shortage of well-trained specialists in this field might well exist.

Molecular biologists currently isolate extremely small amounts of RNA or DNA that code for proteins or enzymes within cells or organisms. Accurate localization of a specific nucleotide sequence frequently depends upon the availability of sequence information. The protein sequence is translated into a nucleotide sequence which is synthesized and subsequently employed in hybridization studies, that is identification of that piece(s) of DNA (or RNA) coding for the desired protein product. The subsequent isolation, cloning, sequencing and expression of the coding sequence then produces the desired protein or peptide product. This simplified overview is the basis of the growing biotechnology industry.

Central to this identification process is the successful isolation and characterization of proteins. This characterization involves not only purification to homogeneity and primary sequence determination, but also the localization, and frequently structural determination, of the positions of post-translational modification. More than 100 different modifications are now recognized. As the repertoire of analytical techniques and instrument capabilities improve, the protein chemist will clearly discover more.

As more complex or poorly expressed peptides and proteins began to be characterized, a serious need for improved isolation and chemical methodologies has developed. For the most part, the proteins considered 'interesting' today are available only in small quantities, often at the microgram level, from sources which are either difficult to obtain in large quantities or prohibitively expensive.

High-performance liquid chromatography (HPLC) and gel electrophoresis are the two techniques which have found extensive use in both the initial isolation and characterization processes of these substances.

To isolate samples in such small amounts, new methodologies and increased detection sensitivities were needed, but not at the expense of significant chemical alteration of the desired sample. Amino-terminal blockage, proteolysis, or amino acid side-chain alteration, for example, the addition or removal of constituents are unacceptable. Consequently, exacting requirements were placed both on the chemicals which directly 'contacted' the sample on the 'wetted' surfaces of instruments, and on the recovery characteristics from surfaces. From these demands, a growing industry for the production

Table 1. Micro-chromatography and gel electrophoresis application areas in protein isolation/characterization.

Purification
Purity estimation
Sample concentration/de-salting
Assessing extent of chemical and enzymatic cleavages
Protein/peptide comparative mapping
Sequencing
Amino acid analysis

Table 2. Physiochemical properties of proteins affecting isolation.

Separation method	*Physical property*				
	Charge	*Hydrophobicity*	*Shape*	*Size*	*Solubility*
Centrifugation			X	X	
Chromatography	X	X	X	X	X
Electrophoresis	X		X	X	
Extraction					X
Gel filtration			X	X	

of high-purity chemicals and 'inert' instrumentation has arisen. *Table 1* lists the general applications in which either HPLC or electrophoresis, or both, prove useful to the protein chemist. These methods have been optimized for analytical and preparative applications on smaller and smaller amounts of material.

The physiochemical properties of samples which enable one to differentiate and thereby separate proteins and peptides are compared in *Table 2*. These properties also form the basis upon which most characterization techniques are dependent. Of the techniques listed in *Table 2*, chromatography is the only one which utilizes each of the different physical properties. The solubility differences, for example, among proteins in the various reagents and solvents used during Edman sequencing prevent them from washing out. Similarly, hydrophobic interaction chromatography relies on the 'salting-in and -out' of proteins in high-versus-low salt concentrations.

Ion-exchange columns separate on the basis of charge; hydrophobic differences are the basis of reverse-phase separations. The size exclusion packings differentiate by molecular shape and size and hydrophobic interaction packings separate by solubility. Electrophoresis achieves its characteristically high resolution through charge and/or size differences. The use of these two separate techniques, HPLC and electrophoresis, either singly or in combination, currently provide the optimal means of isolating small quantities of polypeptides.

Before proceeding to the following narrative, a few definitions are appropriate. 'Micro'-levels are defined as $1-5$ μg for proteins and at the microgram, or less, for peptides. These quantities correspond to the subnanomole level, and frequently the tens of picomoles, for proteins in the 50-kd (or under) molecular weight range. Currently, state-of-the-art instrumentation allows even those researchers with limited experience to determine extended ($20-40$ residues) amino acid sequences on such small amounts. Isolating, as well as handling, such minute quantities requires, however, specialized knowledge.

In the following text, where 'recovered by lyophilization' is mentioned, it implies that a liquid sample has been lyophilized under vacuum (<100 mTorr) while centrifugation was simultaneously performed. The sample might have been frozen before vacuum introduction, depending on the volume and nature of the solvent. Samples containing relatively concentrated acids, and some organic solvents, require dilution with water prior to freezing.

Whenever possible, liquid samples are collected into polypropylene tubes (Eppendorf™, or the like) and stored frozen at −20°C until use. When samples remain frozen under these conditions oxidation reactions are reduced, sample losses minimized, concentrations remain known (assuming zero evaporation) and re-solubilization is not a potential problem. Sample storage in a dried/lyophilized state is not encouraged. Samples recovered from gels by electroblotting onto the appropriate glass surfaces should be stored at −20°C in an inert atmosphere, argon or nitrogen, minimally 99.998% pure.

2. MICROBORE HPLC INSTRUMENTATION

2.1 **Hardware considerations**

The need to develop techniques for micro-isolation required increased detection sensitivities and then the appropriate sequencing techniques with which to characterize an isolate. The preparation of a few micrograms, rather than tens or hundreds of

Table 3. HPLC hardware considerations in narrow and microbore chromatography.

Pumps	Variable flows, 1−1000 μl/min
	Flow-rate resolution, 1 μl/min
	High compositional accuracies (<2% RSD) and minimum pulsation (<5 p.s.i.)
	Operation independent of solvent viscosity and compressibility
	Pressure limits of 2000−5000 p.s.i.
Gradients	Control at the 1-μl level
	Compositional accuracy ±1%
	Minimized gradient distortion during dynamic mixing
	Matched mixer volume for desired sensitivities
Injectors	Volume selectable through loop changes (μl to ml)
	Zero dead-volume without cross-contamination
	Accuracy and precision maximized
Columns	Commonly stainless-steel with polished, regular inner surfaces
	Zero dead-volume end assemblies
	Stainless-steel frits with minimized absorption characteristics
	Simplified column-to-column change, cartridge design optimal
Connections	Corrosion-resistant, high tolerances on diameters and concentricity
	Appropriately end-cut and cleaned
	Minimized band broadening or extra-column dispersion
	Stainless steel nuts, ferrules and unions; alternatively, finger-tight fittings of Kel-F or another inert material
Solvents	High purity−low background, low levels of trace impurities, minimized residue and particulate content
	Inert or non-reactive with LC components and chromatographed substances
	Appropriate solvent strength, UV cut-off, boiling point and viscosity
	Minimal toxicity and flammability

micrograms, from liquids (cell culture supernatants, urine, fermentation media) or solids (tumours, tissue extracts) is currently achievable within hours, not days or weeks. The cost of the starting material and the time required for work-up have been proportionately reduced. Chromatography, in particular HPLC, is the one technique readily adaptable to these varied prerequisites.

HPLC components are specifically designed and optimized to achieve the necessary performance. The important hardware considerations are listed in *Table 3*.

Because gradient elution and small diameter columns are basic requirements for successful HPLC isolations, the delivery of liquids in precise volumes and compositions is of paramount importance (1,2). Not only must the 'dead' volume of the system be minimized (3,4), but the mixing of the two (or more) requisite solvents/buffers must be efficient (5). An optimized mixing method that filters 'noise' from the pumps at very high sensitivities is required. Injectors for analytical and preparative analysis at micro-levels are most often manually operated. Fraction collection is also performed manually because peak volumes are commonly less than 100 μl. While these operations could be automated, current instrumentation has not been optimized for minimal 'dead' volumes, to maximize the percentage of sample injected, or to collect microlitre amounts into relatively small vials. Thus, neither a 'smart' fraction collector capable of peak detection and isolation nor a miniaturized autosampler is currently available.

2.2 Column parameters

The important column variables are listed in *Table 4*. Physical size (inner diameter and length), the support material and its chemically modified surface and the conditions used for chromatography are given (6,7). As will be explained later, microbore (1 mm i.d.) and narrowbore (~2 mm i.d.) columns of minimal lengths are employed for most applications (8–11).

Peptide and amino acid derivative HPLC is still optimally performed on silica supports. These supports are, however, now often replaced by polymeric materials. Isolations, for example, requiring buffers of higher pH (>8) or minimized denaturation and subsequently higher activity yields, are often conducted on these substitutes.

Column supports with particle sizes of 5–7 μm, 300 Å pore sizes, and either C4 or C8 surface bondings seem equally applicable (12,13). While maintaining reasonable

Table 4. Most commonly used column parameters for protein/peptide HPLC.

Parameters	*Ranges*
Columns size	
Diameter (mm)	1–4.6
Length (mm)	1.5–250
Column supports	
Particle size (μm)	5–20
Pore size (Å)	100–300
Surface bonding	Varies with chromatographic mode
Separation conditions	
Temperature range (°C)	5–40
Flow-rates (μl)	10–2000

Table 5. Most commonly used buffers for protein/peptide HPLC

Reverse-phase	*Acidic pH (2–4)*	*Neutral pH (4–7)*
	0.1% TEA, HFBA or H_3PO_4	10–100 mM NH_4, TFA or sodium acetate
	5–60% formic acid	50–100 mM NH_4 or NaH_2PO_4
		10–50 mM Tris, NH_4HCO_3 or KH_2PO_4
	Acetonitrile, propanol or some organic mixtures are used as the eluant.	
Ion-exchange	*Buffer (10–50 mM)*	*Salts (up to 1 M)*
anion	Tris, bis-Tris or phosphate	Chloride, acetate or phosphate
cation	Sodium acetate or phosphate	Sodium chloride
Hydrophobic	*Buffer (0.1–0.5 M pH 6–7)*	*Salts (up to 3 M)*
	Ammonium acetate	Ammonium acetate
	Sodium phosphate	Ammonium sulphate
	Potassium phosphate	Sodium sulphate

Abbreviations: HFBA—heptafluorobutyric acid; TFA—trifluoroacetic acid.

operating pressures, these column parameters optimize recoveries, roughly 50 and 80% for proteins and peptides, respectively.

Recently, some construction materials in chromatographic equipment have come under suspicion for chemically altering the samples. The end frits that secure the support material in the column have been reported to absorb proteinaceous material and to release it either slowly or not at all (14–16). Trace components, such as certain metal ions, are also known to leach out of stainless steel frits, tubing and wetted surfaces of the pump under conditions of high salts and/or extremes of pH. Documented effects of these trace compounds on isolates are lacking, but researchers should be aware of the possible phenomenon. Given the mild conditions which are normally employed for an isolation, one should probably first suspect the trace metal contaminants in water, solvents and/or buffers.

2.3 Buffer systems

Table 5 describes the buffer combinations employed for the three most useful chromatographic modes. The choice of one or another mode depends upon the desired application and properties of the sample. For example, the use of a low pH buffer system and an organic eluant for the isolation of an enzyme that is denatured under such conditions is ludicrous. Simple incubation experiments in vials will indicate the appropriateness of a chosen set of conditions.

Mass recoveries can only be estimated by chromatography with preliminary studies on small quantities. Frequently these results are misleading because recoveries often improve when larger amounts are chromatographed.

2.4 Detection enhancement

To improve detection for an application one may increase detector or recorder sensitivity settings, modify the proteins with more easily detectable reporter groups or simply reduce column inner diameter and flow-rate.

Reducing the column diameter to 2 mm, an adjustment possible with most modern HPLC equipment, and flow by a factor of five increases mass sensitivity by 5-fold.

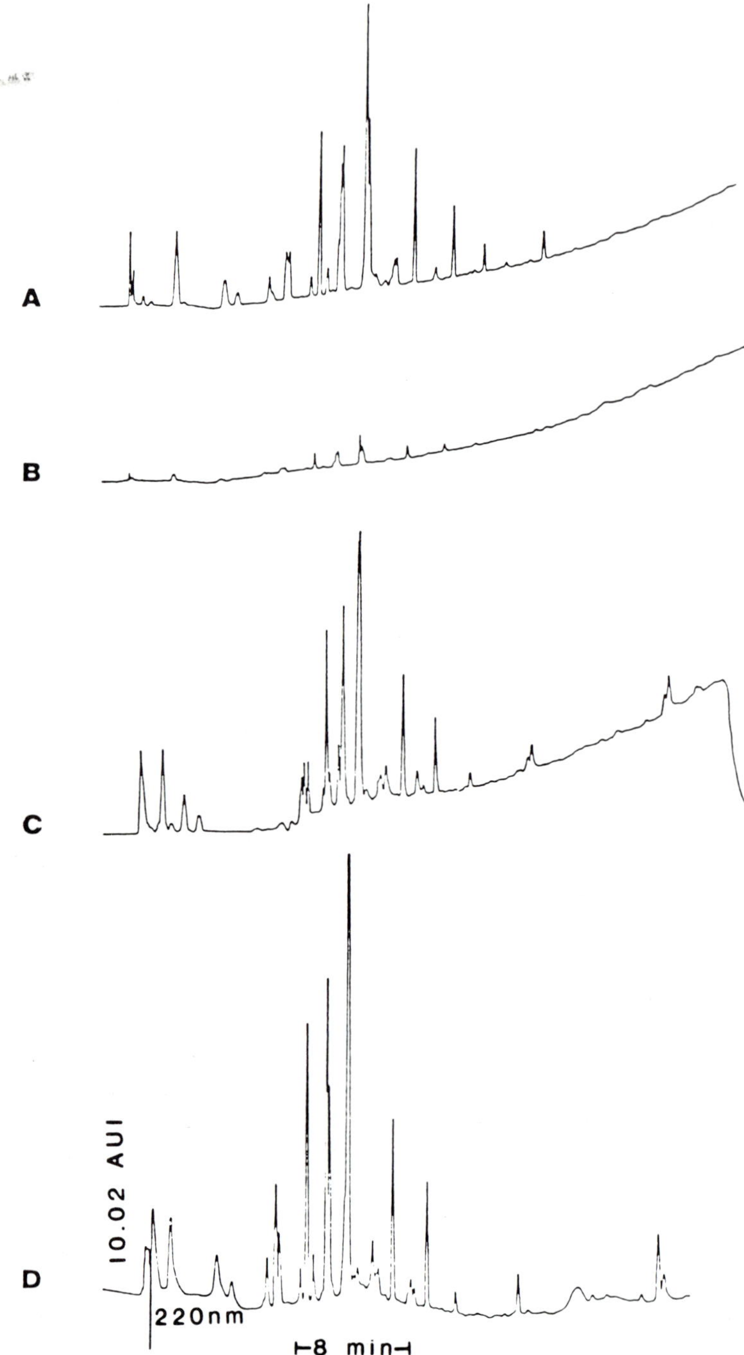

Figure 1. Sensitivity enhancements of narrower column diameters. A tryptic digest of apomyoglobin was chromatographed on a series of reverse-phase columns packed with an identical support (Aquapore™ RP-300) using a 45-min linear gradient from 0% B buffer (0.1% TFA) to 100% B (60% A in 0.1% TFA). (**A**) 4.6 × 250 mm at 1 ml/min and 1000 pmol; (**B**) as in **A** except 100 pmol material injected; (**C**) 2.1 × 220 mm at 200 μl/min and 200 pmol; (**D**) 1 × 250 mm at 50 μl and 100 pmol.

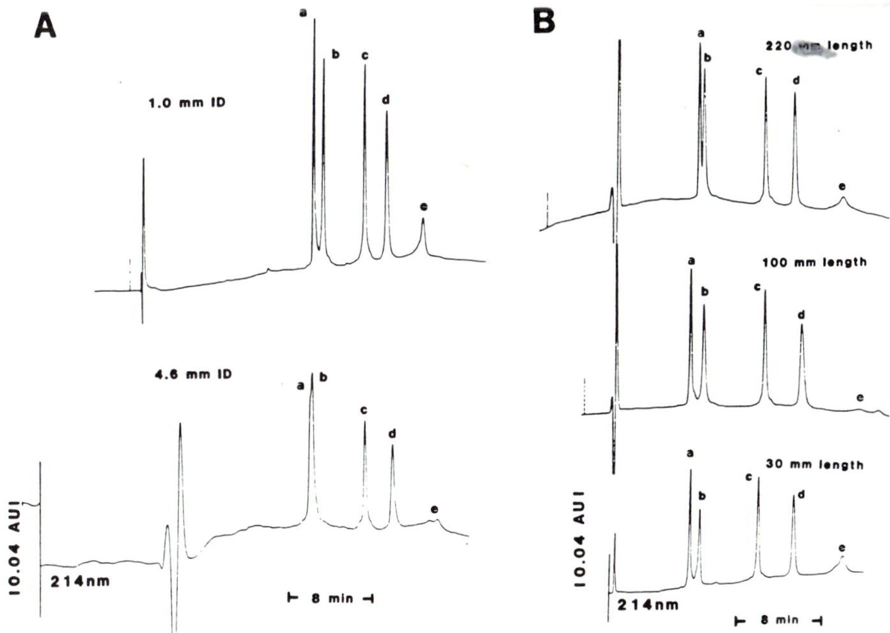

Figure 2. Effects of column diameter (**A**) and length (**B**) on low flow-rate separations. Column sizes were 1 × 250 mm and 4.6 × 220 mm in **A** and 2.1 mm diameters in **B**. All chromatography was performed using the TFA/AN buffer system at 100 μl/min flow-rates. A protein mixture containing insulin (a), cytochrome *c* (b), β-lactalbumin (c), carbonic anhydrase (d) and ovalbumin (e) was used (cf. ref. 18).

This increased sensitivity is also affected by instrument factors including extra column dispersion, gradient mixer, column efficiency and flow cell design. At very high detection levels, 10–20% differences between instrumentation composed of the same components also appear. Consequently, actual increases in sensitivity must be quantitated for each individual chromatographic system.

Figure 1 illustrates sensitivity enhancements observed for tryptic maps of apomyoglobin. The quantities injected decrease from 1000 pmol on a 4.6 mm i.d. column (*Figure 1A*) to 100 pmol on a 1.0 mm i.d. column (*Figure 1D*). Thus, increasing the mass per unit volume of eluant enhances detectability. This factor is one of the basic requirements for micro-level purification and characterization (17).

It should be clear that pre-column dead volumes, tubing, mixer and the injector only effect the shape or accuracy of the gradient profile, not peak shape or profile. Each, and all, of these factors are, however, extremely important for isocratic elution.

Flow cell volumes are often considered important in microbore HPLC. For gradient elution this is not generally the case; only minor resolution effects are observed between 12 μl and 2.5 μl cell volumes. Light path length is more important. Minor obstructions, reduced lamp output (energy) or poor alignment and/or refractive index effects all combine to set the upper sensitivity limits of a given detector.

Flow cell volume is important when peaks elute in relatively small, closely spaced volumes. The 'dead' volume of the tubing from the cell to the point of collection is also extremely important. If 1 mm i.d. supports are utilized for preparative

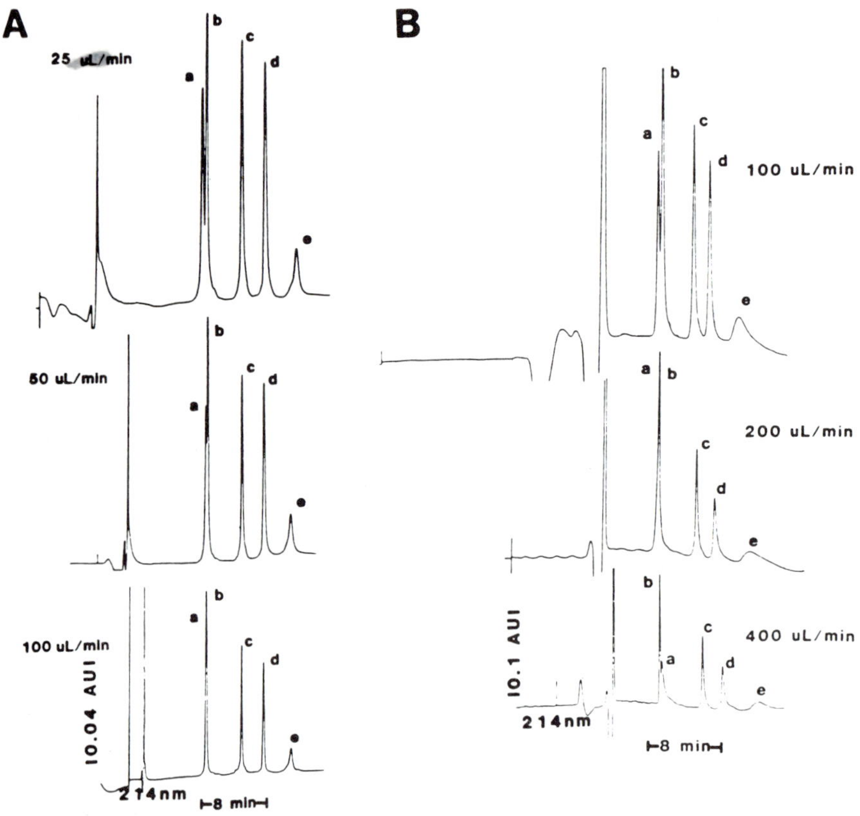

Figure 3. Effect of flow-rate on protein separation. The standard protein mixture (*Figure 2*) was injected onto either 1.0 mm (**A**) or 4.6 mm (**B**) i.d. columns. Elution was performed at the indicated flow-rates using a TFA/AN buffer system (cf. ref. 18).

purposes, then small volume cells (~2.5 μl) with reduced 0.1 mm Teflon i.d., fused silica or stainless steel tubing should be employed. These precautions minimize cross-contamination and peak dilution.

2.5 Effects of column diameter and column length

At relatively low flow-rates protein separations are independent of column length (*Figure 2B*, cf. ref. 18). Column efficiency and/or small differences in organic eluant concentration, however, are important factors at these flow-rates (see insulin–cytochrome *c* separation in *Figure 2A*). Although several investigations (8,10,11) have clearly shown that column length is unimportant in protein HPLC, the results have not been generally accepted. Nonetheless, as peak e in *Figure 2* demonstrates, the losses of hydrophobic proteins, such as ovalbumin or various membrane proteins, increase with increasing column size, that is larger total surface area. If recoveries are important, then smaller diameter, shorter columns should be used. This particular column configuration also reduces operational costs significantly.

2.6 Flow-rate effect on separation

A reduction in the elution rate frequently improves chromatographic separation as *Figure 3* clearly illustrates (cf. ref. 18). This improvement is assumed to be caused by slight concentration differences in the organic eluant as flow-rates vary. Thus for a given column diameter, an optimized set of elution conditions needs to be determined experimentally prior to utilization.

Figure 3 also illustrates the 2- to 3-fold increase in sensitivity one can achieve by reducing the flow-rate by a factor of 4. When utilizing 1 mm i.d. supports reducing extra-column effects are extremely important and should be kept at a minimum.

2.7 Flow-rate effect on peak elution volume

The overall linear relationship between flow-rate and peak volume is relatively independent of column length, as illustrated in *Figure 4*. A 4-fold reduction in flow-rate produces approximately a 3-fold decrease in peak volume. Because the glass fibre filters used in gas-phase sequencing can only retain approximately 30 μl of liquid, flow-rates must be optimized for each column diameter to produce appropriate peak volumes. Utilizing the smallest column diameter, for example 1 mm i.d., easily reduces the total volume of a collected peak to approximately 30 μl.

3. ELECTROPHORESIS EQUIPMENT

3.1 General concerns of chemical modification

Proteins can be modified chemically during purification. The investigator must constantly review the quality of his chemicals, instrumentation and sample recovery methods to ensure that such chemical changes are minimized, or eliminated. Unfortunately the measurement, or even estimation, of protein modification is extremely difficult to conduct accurately.

One can easily determine initial sequencing yields before and after a particular chemical derivatization or separation step. The quantitative value of such determinations, however, depends on numerous factors. These include sample-to-sample reproducibility of the sequencer chemistry, the accuracy of the HPLC separation and quantitation of the phenylthiohydantoin (PTH) amino acid derivatives, one's ability to dissolve and transfer samples quantitatively and, of course, the reproducibility of the initial chemical/separation treatment itself.

There is a major drawback. The Edman chemistry can only estimate, when performed correctly, the degree of N-terminal blockage and/or modification of amino acid side-chains located within the N-terminal region of the sample. Any modifications that occur elsewhere in the molecule will therefore go undetected by this method. Other means must then be used for comparing samples before and after a given isolation step. Electrophoretic separations in various media, isoelectric focusing and chromatofocusing can all detect changes in charge characteristics. Similarly, molecular weight alterations can be observed by size-exclusion chromatography or electrophoresis under denaturing conditions. If modifications have occurred which do not affect either charge or molecular weight, then their detection becomes extremely difficult.

Polyacrylamide gel electrophoresis (PAGE) is the second most widely used

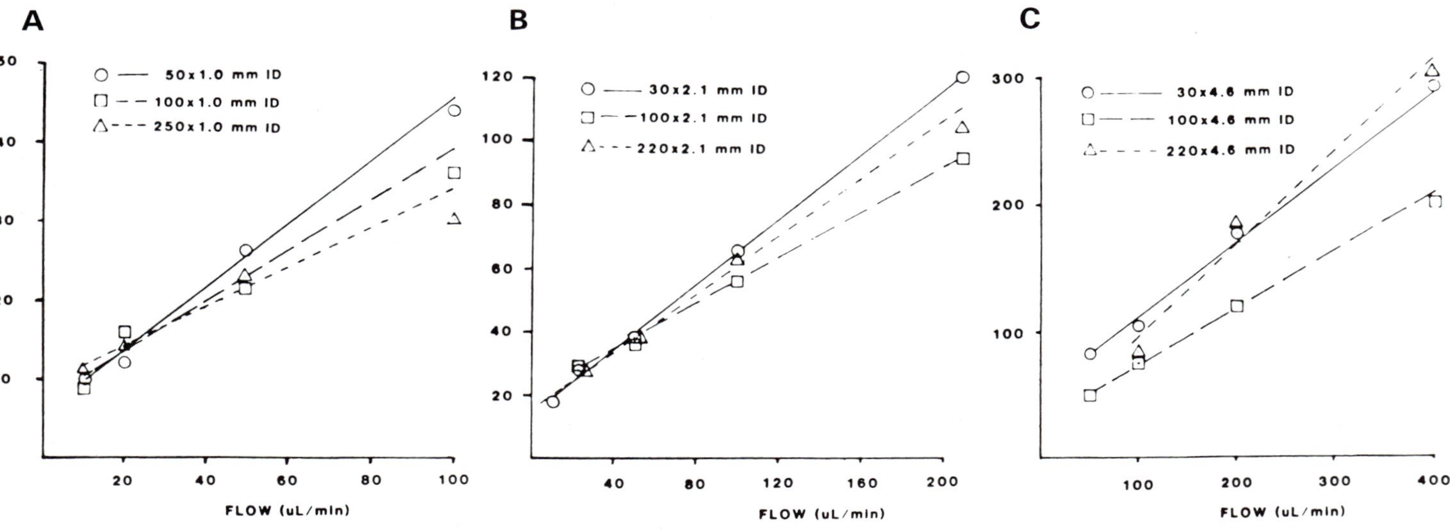

Figure 4. Peak elution volumes as a function of flow-rate and column size. Five proteins (cytochrome *c*, insulin, β-lactalbumin, carbonic anhydrase and ovalbumin) were chromatographed on 4.6 mm (**A**), 2.1 mm (**B**) and 1.0 mm (**C**) i.d. columns at the indicated flow-rates. Their peaks were averaged and plotted as illustrated. See ref. 18 for further details.

Table 6. General concerns for protein samples recovered from PAGE for sequence analysis.

Problems	*Solutions*
Amino-terminal blockage and/or side chain modification	1. Use high-purity gel chemicals 2. Gel ageing prior to sample separation 3. Mini-gel system used to maintain high protein-to-gel ratio 4. Use of free radical and oxidant scavengers during electrophoresis
Protein fragmentation	1. Shorter staining periods 2. De-staining at 4°C 3. Denature samples prior to electrophoresis (minimizes proteolysis)
Contaminant introduction	1. Electrodialysis after electroelution 2. Organic solvent precipitation

micro-isolation method. *Table 6* lists the possible modifications that can occur during each phase of electrophoresis: sample work-up, separation, staining–de-staining or recovery. One should carefully check each procedure independently, then collectively to identify problems.

Electrophoresis requires a large number of chemicals, each of the highest purity. REMEMBER that this doesn't necessarily mean highest chemical purity, but rather an inertness purity, or absence of undesirable modification activity. Gel ageing and pre-running can reduce the presence of unpolymerized materials (acrylamide etc.) and the catalyst (ammonium persulphate), which might cause problems. Similarly, the protein-to-gel ratio is maintained as high as possible using mini-gel systems which, in turn, are run in the presence of free radical scavengers.

Protein fragmentations also occur that alter native structures. This is most likely to happen during either sample preparation or detection and recovery following electrophoresis. Simply the re-electrophoresis of a small portion, usually 5–10%, of the recovered sample will indicate whether fragmentation has occurred.

Although often not possible to completely circumvent, one must minimize contamination. One approved method uses an electrodialysis step in dilute buffers which reduces salt, denaturant and buffer concentrations. Following lyophilization, the remaining contaminants can be reduced further by an organic precipitation procedure.

3.2 Reagents

The list of reagents given in *Table 7* represents the chemical sources that we use for preparing sodium dodecyl sulphate (SDS) gels for sample isolation as described in ref. 19. We cannot overemphasize that the inertness of many of these chemicals must be confirmed using protein standards and sequencing prior to undertaking the isolation and characterization of unknowns.

3.3 Equipment

Electrophoresis equipment for microscale isolation is described in *Table 8*. Separations are performed on vertical mini-slab gels of thicknesses 1 mm or less prior to

Table 7. Reagents used for preparing SDS gels.

Item	*Source*	*Cat. no.*
3-(Cyclohexylamino)-1-propanesulphonic acid	Aldrich	16,376-7
Cyanogen bromide	Aldrich	C9,149-2
Formic acid	Aldrich	25.136.4
Methanol	Aldrich	27.047.4
Polyvinyl pyrrolidone (PVD-40)	Aldrich	85,656-8
Acetonitrile	Applied Biosystems	400313
Trifluoroacetic acid	Applied Biosystems	400003
Filter paper	Am. Sci. Products	F2413-24
Glass fibre sheet	Am. Sci. Products	F2834-125
Petri dishes	Am. Sci. Products	D2005-6
Acetic acid	J.T.Baker	9515-3
Sodium bicarbonate	J.T.Baker	3506
Glycerol	BRL	5514UA
Acrylamide	Bio-Rad	161-0100
Ammonium persulphate	Bio-Rad	161-0700
Bromophenol blue	Bio-Rad	161-0404
Glycine	Bio-Rad	161-0717
N,N'-Methylene bis-acrylamide	Bio-Rad	161-0200
Sodium dodecyl sulphate	Bio-Rad	161-0301
N,N,N,N-Tetramethylenediamine	Bio-Rad	161-0800
Tris (hydroxymethyl) aminomethane	Bio-Rad	161-0716
Dithiothreitol	Calbiochem	233153
Polybase filter	Janssen	24-394-47
Polyvinylidene difluoride (PVDF)	Immobilon, Millipore	1PVH304FO
3,3-Dipentyloxacarbocyanine iodide	Molecular Probes	D-273
N-Trimethoxysilylpropyl *N,N,N*-trimethylammonium chloride	Petrarch Systems	T-2925
3-Aminopropyltriethoxysilane	Petrarch Systems	A0750
N-Ethylmorpholine	Pierce	20805
Guanidine–HCl (8 M)	Pierce	24115
Heptafluorobutyric acid	Pierce	25003
Coomassie blue R-250	Sigma	B0630
Glutathione	Sigma	G4251
2-Mercaptoethanol	Sigma	M-6250
Ponceau S	Sigma	P7767
Sodium thioglycolate	Sigma	T-0632
Tween 20	Sigma	P1379
Trypsin	Worthington	77617m

sample recovery. If larger amounts, in the milligram range, are isolated, the electroelution–electrodialysis procedure (20) developed for active elution of stained protein into a specially designed cell is used. For smaller sample amounts the newly developed electroblotting method onto appropriately prepared glass surfaces is employed (21–23).

The equipment sources used are:

(i) mini SDS–PAGE: Bio-Rad Model 360 mini-vertical slab cell;
(ii) electrophoretic blotting: Bio-Rad trans-blot electrophoretic transfer cell;
(iii) UV lamp: model UV6L-25 Minerallight Lamp;
(iv) power supply: Bio-Rad Model 250/2.5 constant voltage supply, Buchler 3-1500 constant power supply;

Table 8. Electrophoresis apparatus for micro-scale protein sample preparation.

Slab gel	Mini vertical Studier-type slab gel system (10 × 10 cm) Gel thickness ≤1 mm
Electro-eluter/concentrator	Three-chambered electrophoresis tank[a] with a two-channel peristaltic pump for buffer recirculation Two-chambered electrophoresis tank[b]
Electroblotting cell	Gel-holder–cassette type system Low voltage-high current and high-transfer field strength Mini system if available
Power supply	Constant voltage adjustable to 0–250 V, 0–250 mA Overload and short-circuit protection

[a]Designed by Hunkapiller *et al.* (20) and available for purchase through Division of Biology, California Institute of Technology, Pasadena, CA 91125, USA.
[b]From C.B.S. Scientific Company, Del Mar, CA 92014, USA.

(v) shaker: Lab-Line: Orbit shaker.

3.4 Preparation of gel solutions, buffers and slab gels

Some precautions are necessary when preparing the gels and samples for electrophoresis. Perhaps foremost is the use of disposable plastic gloves which minimize contamination and prevent direct physical contact with the solutions. Acrylamide and bis-acrylamide are neurotoxins, as well as apparently carcinogenic. Contact should therefore be minimized. All glassware used to prepare or store solutions should be freed of possible contamination. Similarly, glassware and pipettes or pipette tips that are used for mixing and/or preparing the gel solutions need to be clean.

Of the chemicals listed in *Table 7*, only SDS must be re-crystallized (twice) from ethanol–water using the method described in ref. 20. One should filter those stock solutions indicated with an asterisk below prior to use (Millipore filter, HVLP02500).

3.4.1 *Stock solutions and buffers*

(i) *Lower Tris (4 ×).* Dissolve 36.34 g of Tris base and 0.8 g of SDS in 150 ml of de-ionized water, and titrate the pH to 8.8 with 6 M HCl; add de-ionized water to a final volume of 200 ml.

(ii) **Upper Tris (4 ×).* Dissolve 12.11 g of Tris base and 0.8 g of SDS in 150 ml of de-ionized water, and titrate the pH to 6.8 with 6 M HCl; add de-ionized water to a final volume of 200 ml.

(iii) **30% acrylamide.* Dissolve 30 g of acrylamide and 0.8 g of bis-acrylamide in de-ionized water to a final volume of 100 ml.

(iv) *Electrophoresis buffer.* Dissolve 3.03 g of Tris base, 14.41 g of glycine, and 1 g of SDS in 1 litre of de-ionized water.

(v) *10% ammonium persulphate.* Dissolve 100 mg of ammonium persulphate in 1 ml of de-ionized water. Store the solution at 5°C and discard after 1 week.

(vi) *Sample preparation solution.* Mix 1 ml of glycerol, 0.5 ml of 2-mercaptoethanol, 0.3 g of SDS and 1.25 ml of Upper Tris (4 ×) buffer in de-ionized water to make a final volume of 10 ml.

3.4.2 *Gel solutions*

Stock solutions	5% Stacking gel (2 ml)	10% Resolving gel (4 ml)	15% Resolving gel (4 ml)
30% Acrylamide	0.33 ml	1.33 ml	2.0 ml
Upper Tris (4 ×)	0.50 ml	–	–
Lower Tris (4 ×)	–	1.0 ml	1.0 ml
De-ionized water	1.17 ml	1.67 ml	1.0 ml
	De-gas for 5 min		
TEMED	2 μl	2 μl	2 μl
10% Ammonium persulphate	15 μl	15 μl	15 μl

3.4.3 *Slab gel preparation*

Prepare gels by following the recognized procedures (24).

3.5 Sample preparation, loading and electrophoresis

The physical nature of a sample determines the preparation procedure for electrophoresis. Because only limited sample volumes can be applied to mini-slab gels, samples are normally concentrated and/or lyophilized prior to the dissolution in or addition of the electrophoresis sample buffer. Sample volumes from 10 μl and 100–200 μl have been found quite ample for analytical and preparative purposes.

If contaminating salts or other substances are present which might interfere with the separation, one must remove them before final solubilization. Dialysis, salt or trichloroacetic acid (TCA) precipitation, and/or membrane concentrations should not be considered because losses at the microgram level are frequently substantial. Short reverse-phase supports are exceptionally good for fast de-salting and concentration by HPLC (see Section 4.2). Collection into vials of limited volume (500 or 1500 μl microcentrifuge tubes) and subsequent lyophilization leaves the product ready for analysis.

(i) To prepare the sample for application, mix equal volumes (1–10 μl) of a liquid sample and sample preparation solution, or add an appropriate volume of 1:1 dilution of sample preparation solution to the lyophilized product.
(ii) Heat in boiling water for 2 min.
(iii) Add 1–3 μl of 0.1% bromophenol blue and load onto the gel within 30 min.
(iv) Pre-run the SDS gels by adding 350 μl of a reduced glutathione solution (10 mM, stored under argon at –20°C) per 70 ml electrophoresis buffer to the upper reservoir and apply a 3 mA constant current for 2 h.

(v) Decant the pre-run buffer, refill the reservoirs with fresh buffer solutions, and load the samples into the appropriate wells.
(vi) Add sodium thioglycolate (100 mM, 70 μl per 70 ml) to the upper reservoir and perform electrophoresis at 7 mA constant current for 1 h.

Sample recovery from the gel presents the next challenge and, as already emphasized, another procedure that can modify the desired product. Two methods are currently employed and each is reviewed in the following two sections.

3.6 **Electroelution–electrodialysis recovery method (20)**

3.6.1 *Preparation of dialysis membrane*

Spectrapor membrane is available from Spectrum Medical Industries with molecular weight cut-offs of 3500–50 000. This is cut into 6 inch lengths and soaked in 1% $NaHCO_3$ at 60°C for 1 h, then 0.1% SDS for 1 h then finally washed well with water. Store in 0.1% SDS, 0.1% NaN_3 at room temperature.

3.6.2 *Electrophoretic elution apparatus*

This consists essentially of 'H'-shaped elution and concentration chambers linked by a cross channel and closed off from the electrode buffers by dialysis membrane. Home-built equipment which works well is described in ref. 20. Commercial apparatus is also available (e.g. from C.B.S. Sci. Co., Del Mar, CA 92014). Use discs of Spectrapor membrane with a molecular weight cut-off just lower than the protein of interest.

3.6.3 *Preparation of samples for electroelution*

(i) Stain the SDS–PAGE gel as described in Section 3.5 with 0.5% Coomassie brilliant blue in freshly prepared acetic acid:isopropanol:water (1:3:6 by vol.) for 15 min using agitation.
(ii) De-stain in acetic acid:methanol:water (50:165:785 by vol.) until bands are visible. Laboratory tissues or ion-exchange beads will speed de-staining.
(iii) Cut out the bands on a light box and soak in several changes of water over 2 h. At this stage the bands can be stored at −20°C.
(iv) Dice the gel slices into 1 mm cubes under water using a safety razor.
(v) Aspirate the water and soak the pieces in elution buffer (0.1% SDS, 0.05 M NH_4HCO_3) for 5 min.
(vi) Remove the elution buffer, load the gel fragments into the larger chamber of the electroelution apparatus and just cover them with soaking buffer [2% SDS in 0.4 M NH_4HCO_3 and stock 10% dithiothreitol (DTT) to a final concentration of 0.1% DTT].
(vii) Overlay the soaking buffer with elution buffer until this is continuous between the two chambers.
(viii) Assemble the apparatus and allow the gel fragments to soak for 3–5 h.
(ix) Fill the electrode chambers with elution buffer and arrange for re-circulation and mixing of the two electrode buffers (3 ml/min). The gel loading well should be nearest the cathode.

(x) Run the elution at 50 V for 12–16 h.

(xi) After this time replace the elution buffer in the electrode chamber with dialysis buffer (0.2% SDS in 0.01 M NH_4HCO_3) and continue electrodialysis at 80 V for 20–24 h.

(xii) At completion of the dialysis step, the sample should be concentrated as a narrow, approximately 1–2 mm blue band on the cathode side of the elution cell. If the Coomassie blue is more diffuse, it can be concentrated by replacing the 'used' electrodialysis buffer with a freshly prepared solution and again electrophoresing for 30–90 min.

(xiii) Carefully remove the colourless buffer from both sides of the elution cell, leaving the undisturbed blue concentrate in a minimal volume of approximately 100 μl.

(xiv) Mix the concentrate, remove with either a Pasteur pipette or a glass syringe, and transfer into a 1.5 ml minicentrifuge tube.

(xv) Carry out multiple washings with distilled water to assure complete transfer.

(xvi) Cover the vial containing approximately 500 μl of liquid with Parafilm.

(xvii) Make a number of perforations in the Parafilm with an appropriately sized needle, freeze the sample and lyophilize.

(xviii) After removing the Parafilm and capping, samples can be stored at −20°C or processed further for SDS–salt removal by precipitation (see Section 3.8.1).

3.7 **Electroblotting, glass fibre derivatives (22)**

Prior to the electrophoretic separation on a slab gel, the glass blotting filter derivatives need to be prepared.

3.7.1 *Activation of glass fibre sheets (19)*

(i) Slide six sheets of Whatman GF-F filter into a glass Petri dish containing 60 ml of neat trifluoroacetic acid (TFA). Cover with the second half of the dish. Keep the filters in acid for 1 h at room temperature. (**Note**: remember that this and the next step must be performed in a fume hood.)

(ii) Decant the acid and place each sheet on top of a few layers of paper towelling. Air dry for 1 h.

(iii) Place the filters in a vacuum desiccator and dry overnight under vacuum (~100 mTorr).

3.7.2 *Glass fibre sheet derivatization*

(i) Prepare the silane solutions by mixing in a glass Petri dish 2 ml of 3-aminopropyl-triethoxysilane (AP), or *N*-trimethyoxysilylpropyl-*N,N,N*-trimethylammonium chloride (QA), with 97 ml of acetonitrile and 1 ml of water.

(ii) Slide six sheets of TFA-treated GF-F filters into the solution and place on an orbital shaker. Gently shake for 2 min.

(iii) Decant the silane solution and replace with 100 ml of acetonitrile. Continue shaking for 10 min, then decant.

(iv) Repeat the acetonitrile rinse 10 times as described in step (iii).

(v) Place the rinsed filters in an oven at 100°C for 60 min to cure the silane linkage.

(vi) Store the derivatized filters in a well-covered Petri dish at room temperature in the dark.

3.7.3 *Electroblotting*

(i) Remove the mini-slab gel from the cell, place in a Petri dish, and soak in the electroblotting buffer for 10 min. The electroblotting buffer is a 10-fold dilution of a 0.25 M stock solution prepared by diluting 31.4 ml *N*-ethylmorpholine in 900 ml water, titrating to pH 8.3 with formic acid, and bring the solution to a final volume of 1 litre. This solution is stable for 2 weeks if stored at 5°C.

(ii) Wet the sponges, the Whatman No. 3 filter sheets (same size as the sponge, 20 × 15 m), and the derivatized glass fibre sheets with electroblotting buffer in a suitable container (27 × 20 × 7 cm plastic box).

(iii) Assemble the transblotting sandwich in the following order: anode side sponge, Whatman No. 3 filter, two sheets of derivatized GF-F filters, mini-gel, Whatman No. 3 filter, cathode side sponge.

(iv) Pour 2 litres of electroblotting buffer into the transblot cell and insert the transblotting sandwich.

(v) Place the transblot cover on the unit and begin electroblotting at 50 V (140–170 mA) for 1 h at room temperature.

3.7.4 *Detection*

(i) Remove the derivatized glass fibre sheets from the transblotting sandwich. Insert each sheet between two sheets of dry Whatman No. 3 filter paper and press dry for 30 sec.

(ii) Prepare 30 ml of staining solution and pour into a Petri dish. Staining solution: dissolve 3 mg of 3,3′-dipentyloxacarbocyanine iodide in 3 ml of methanol, then add 27 ml of 50 mM $NaHCO_3$, pH 8.2.

(iii) Stain each blotted glass fibre sheet in the above solution by gently swirling for 2 min. The sheets should be placed into the solution so that the surfaces are evenly wetted.

(iv) Decant the staining solution and rinse with 30 ml of 50 mM $NaHCO_3$ for a couple of minutes then decant the rinse solution. Protein bands of more than 1 μg can be visualized as orange bands against a yellowish background or under short wavelength UV light the bands fluoresce a green–yellow. Detection limits are around 100 ng.

(v) Remove the blotted protein bands using a sharp razor blade and store in a labelled polypropylene microcentrifuge tube.

3.8 **Electroblotting, polyvinylidene difluoride (PVDF)**

3.8.1 *Electroblotting*

The procedures for electroblotting onto PVDF membrane described by Matsudaira (23) can be further elaborated as follows.

(i) Prepare CAPS buffer.
(a) 10 × stock (100 mM, pH 11). Dissolve 22.13 g of 3-[cyclohexylamino]-

1-propane-sulphonic acid in 990 ml of de-ionized water. Titrate with 2 M NaOH (~ 15 ml) to pH 11, and add de-ionized water to make a final volume of 1 litre. Store at room temperature.

(b) Electroblotting buffer. Prepare 2 litres of buffer by mixing 200 ml of 10 × stock buffer, 200 ml of methanol and 1600 ml of de-ionized water.

(ii) Wet the PVDF membranes (2 sheets) with methanol for a few seconds, and place them in a Petri dish containing blotting buffer.

(iii) Remove the gel from the electrophoresis cell and soak in electroblotting buffer for 5 min.

(iv) Assemble the transblotting sandwich, and electroblot at 90 V (300 mA), room temperature for 10–30 min.

(v) Remove the membranes and rinse with de-ionized water prior to the staining.

3.8.2 *Detection*

Protein blotted onto PVDF membranes can be visualized by two different staining procedures with either Coomassie blue or Ponceau S.

(i) Following electroblotting the membrane needs to be rinsed in de-ionized water for 5 min.

(ii) Coomassie blue R-250 (0.1% solution in 50% methanol) staining is carried out for 5 min followed by de-staining (5–10 min) in an acidic methanol solution (5:10:40, methanol:acetic acid:water by vol.).

(iii) Ponceau S (0.2% solution in 1% acetic acid) staining is carried out for 1–2 min followed by a simple rinse with de-ionized water.

(iv) In either case, the stained bands are removed and stored as indicated in step (v) of Section 3.7.4.

3.9 **Sequence analysis**

3.9.1 *Electroeluted–electrodialysed sample*

Samples recovered from SDS gels by the electroelution–electrodialysis method usually need to be freed of contaminating salts and stain prior to sequencing. This is done easily by utilizing a precipitation method that depends on the insolubility of proteins at −20°C in high concentrations of organic solvents. Prior to using this procedure, it is imperative that one check both mass and sequenceable recoveries on protein standards at or near the levels one hopes to perform the recovery of unknowns. Not only can samples be lost by adsorption, but many organics also contain stabilizers that can easily block, in fact quantitatively block, protein amino termini. Thus for this procedure to be successful, one must 'see' the precipitated sample. Access to a 10- to 20-fold source of magnification is essential.

(i) Dissolve the lyophilized plug in 50 μl of distilled water. If flocculent material from lyophilization has collected around the top, it should be solubilized by introducing the water via an appropriate syringe onto the underside of the Parafilm. Shaking or light centrifugation will then transfer the solvent to the bottom of the vial for dissolution of the major portion of the sample.

(ii) Before proceeding further, use magnification to verify that all solid material is in solution. Because of the intense blue colour, this verification involves tilting the centrifuge tube and looking through the liquid in the direction of a strong light. Any solids on the inner wall of the tube should dissolve by gentle vortexing prior to a final centrifugation step which effectively collects the liquid in the bottom of the vial.

(iii) Add 450 μl volume of $-20°C$ ethanol (200 proof, non-denatured), mix the solution and store at $-20°C$ for $4-18$ h.

(iv) Centrifugation pellets the material either at the bottom of the vial (swing-out rotor) or distributes it over a rather large inner surface of the vial (fixed-angle rotor). A smaller distribution area improves the possibility of observing the precipitated material.

(v) At this point the intensity of the Coomassie blue is still approximately the same as it was following electrodialysis and prior to lyophilization. One cannot see the protein precipitate and to do so requires removal of the supernatant with a Pasteur pipette. If all or most of the liquid has been removed and saved in another mini vial, the precipitate is visible at the $1-2$ μg level with magnification. Total removal of supernatant is unnecessary; small amounts of SDS, NH_4HCO_3, ethanol, or stain do not affect either sequencing or PTH identification.

(vi) Re-dissolving the sample is frequently a problem because most of the SDS has been removed by the precipitation procedure. Recognizing that volumes need to be kept at a minimum, add 50 μl of distilled water and check the sample solubility. If insoluble, then introduce 1 μl of a 10% SDS solution and check the solubility again. Because gas-phase sequencing is unaffected by high SDS amounts (≤ 1 mg), multiple additions can be made.

The identification and quantitation of PTH-derivatives are, however, affected by SDS in two ways. When the converted derivatives are collected in a fraction collector, dried, and analysed off-line, the SDS washing out of the sequencer disc contaminates the first couple of tubes. In some HPLC systems this interferes with the subsequent PTH derivative elution and, therefore, separation. For the on-line HPLC system, the SDS will interfere with the transfer step from the conversion flask to the chromatographic unit. This interference manifests itself as hundreds of bubbles in the transfer line which cannot be removed by the limited pressure available to the conversion flask.

In either case, excessive SDS can be removed by performing the following procedure before initiating degradation on a gas-phase sequencer.

(i) Disconnect the transfer line from the cartridge at the A-block and place it in a small vial.

(ii) Deliver the following reagent/solvent combination: (a) R3 for 200 sec, (b) dry, (c) S3 for 200 sec, (d) dry, (e) S2 for 300 sec, and (f) dry.

(iii) Reconnect the transfer line to the A block port and start sequencing.

Overall sample recoveries and molecular weights are conveniently estimated by committing $5-10\%$ of the recovered sample for re-electrophoresis in a mini-SDS gel and staining. A dilution series of standard proteins is also run. Staining the gel using one of the silver procedures detects those proteins present in very small amounts. A Coomassie blue staining of the same gel then detects those present in higher amounts.

Because the latter procedure is more quantitative (staining intensity more linear per mass protein), an estimation made with this dye is therefore more accurate.

3.9.2 *Electroblotted samples*

(i) Cut the blotted bands into 2 × 4-mm pieces and evenly place them into the chamber of the upper cartridge block. **Do not** cut out as 1.2 cm disc because they will affect solvent flow during sequencing.
(ii) If the small pieces are wet, then air dry prior to applying 20 μl of 60 mg/ml polybrene solution.
(iii) Dry again.
(iv) Cover the blotted pieces with a trifluoroacetic acid (TFA)-treated glass filter disc and assemble the cartridge in the normal manner.

Sequencing of samples recovered from polyacrylamide gels by any of the above methods is done using the repair operating programme, for example 03RPTH, or Normal 1.

4. HPLC APPLICATION EXAMPLES

4.1 Comparison of chromatographic methods

Several factors determine the approach that one takes to isolate a given protein or peptide. The physiochemical properties of the sample (see *Table 2*) determine the chromatographic modes that can be used. For example, if the molecular weight is relatively high (40–100 kd) and is sensitive to extremes of pH, then one can select ion exchangers and hydrophobic interaction initially and, perhaps, reverse-phase as the final step. Small peptides are easily isolated using only reverse-phase supports and varying chromatographic conditions.

The ultimate use of the isolated sample frequently dictates the chromatographic mode to use. If activity determinations are necessary then appropriate conditions must be used. It is clear that certain biologically active samples, enzymes, peptides and proteins, undergo reversible denaturation during isolation in acidic media with organic eluants. The regain of *total* native structure is still an open question. Alternatively, other samples are irreversibly denatured under the same conditions.

The relative percentage of the desired sample in the total starting material is similarly important. If, for example, the protein is in urine, in a cell culture supernatant or present in a tissue homogenate then the specific protocols for each would differ. Ion exchangers work well with very dilute solutions; hydrophobic interaction supports are suitable for samples with a high salt content. Reverse-phase columns separate both dilute and high salt samples. Often sample enrichment by precipitation (salt, heat or organic) or a rough molecular sizing step prior to beginning chromatography is necessary. Large amounts of contamination minimize adsorption losses while maximizing recovery due to co-precipitation. Thus, the yields from such enrichments are normally quite high (>95%).

Table 9 illustrates two different approaches to the isolation of human saliva cystatin S, a 14-kd inhibitor whose primary sequence has been determined (25). Two ion-exchange steps were used which required minimally 1–2 days to complete. Our

Table 9. Comparison of the protocols used for isolation and characterization of human cystatin S[a].

	Sample preparation	
Quantity of human saliva	*Isolation procedures*	*Yields (μg)*
4.5 l	*2 × DEAE steps (2.7 × 16 cm)	30 000
50 μl	Single C-8 RP-HPLC step (2.1 × 30 mm)	2
	Amino-terminal sequencing	
Method (amount)		*Number residues identified*
*Manual Edman degradation (not given)		19
Gas-phase sequencer (140 pmol, 2 μg)		50
	Reduction and alkylation	
De-salting/recovery method		*Amount (μg)*
*Sephadex G-75 (1.7 × 90 cm)		30 000
C-8 RP-HPLC (2.1 × 30 mm)		2
	Protein fragmentation	
Separation method		*Amount (μg)*
*DEAE–Sepharose CL-6B (0.9 × 26 cm)		7500
C8-RP-HPLC (2.1 × 30 mm)		2

[a]Data from ref. 25 was used for comparison and designated in the table with an asterisk.

approach employed a single reverse-phase step that lasted less than 1 h. The sensitivities achieved with a narrow bore 2.1-mm i.d. column allowed us to perform the isolation on only 50 μl of initial sample and yielded 2 μg of homogeneous protein. Alternatively, 30 000 μg was purified (25) from 4.5 litres of saliva by a method requiring a number of days. Assuming that the same homogeneity was achieved and that the starting concentrations were identical, the reverse-phase isolation procedure resulted in a 6-fold higher recovery.

Amino-terminal sequencing of the cystatin S was performed on both isolates, 50 residues from 2 μg of our isolate and 19 residues from an ungiven amount in the case of ref. 25. If the determination of only a limited sequence is desired, then a quick 1–2 h isolation scheme yielding small, but workable amounts of sample has advantages. The same micro-methods can be employed to recover samples after chemical modifications, enzymatic or chemical cleavages or product comparison.

4.2 Micro-level alkylation and sample recovery

Some amino acid side-chains require selective modification for amino acid analysis, introduction or limiting of cleavage positions, localization of modified residues or finding and localizing active sites.

Table 10. Optimized protocol for micro-level protein alkylation, specifically S-pyridylethylation.

1. Prepare the following stock solutions and filter prior to use.
   a. 1 M Tris–HCl, pH 8.5, containing 4 mM EDTA
   b. 8 M guanidine–HCl
   c. 1:10 water dilution of 2-mercaptoethanol
2. Mix the stock solutions in a 1-to-3 ratio yielding 6 M concentration of the denaturant buffered with 0.25 M Tris.
3. Dissolve or dilute the sample (1–10 μg) in ≤50 μl of the above diluted denaturant and add 2.5 μl of diluted 2-mercaptoethanol. Incubate at room temperature in the dark under argon for 2 h.
4. Add 2 μl of 4-vinylpyridine (undiluted); mix and incubate as above. **Note:** reagent polymerization is reduced by using recently purchased material that is stored under argon at −20°C.
5. De-salt immediately, by chromatography on a short reverse-phase support. See *Figure 5* for example.

Alkylation is frequently used to modify proteins. As an example, cysteinyl residues modified to yield a stable derivative for hydrolysis and Edman sequencing. One major drawback to alkylation is the difficulty in recovering the modified material. Historically when milligram amounts were involved in this type of research, samples were dialysed extensively, or passed over a size exclusion support and recovered in large volumes requiring lyophilization. In those cases mass losses were minimized by the amounts of material used. At the microgram level, however, such de-salting/recovery techniques frequently lose most of the sample.

A protocol useful for protein alkylation, S-pyridylethylation, at the micro-level is given in *Table 10*. This reaction is preferable because the PE-Cys derivatives are conveniently positioned on either PTC- or PTH-analysers, the derivatives stable to acid hydrolysis and the reaction is quantitative. This protocol maintains small volumes with appropriately high molar concentrations, which are then easy to manipulate for chromatography. For other methods, see Section 6.1, Chapter 6.

Figure 5 illustrates the de-salting of 2 μg pyridylethylated human cystatin S. The major early eluting peaks represent the denaturant, reaction by-products and reagents, while the single designated peak identifies the alkylated product. The recovery volumes from 2.1-mm i.d. columns are sufficiently low, approximately 200 μl, that concentration steps like technique lyophilization, or drying, are not required prior to sequencing or for other characterization procedures.

Although *Figure 5* illustrates de-salting after an alkylation step, it is important to remember that the products from most modifications can be recovered similarly. The same procedure works well for concentrating dilute samples or for changing from one buffer to another. Although use of columns with minimum support volume assures higher recoveries, one must keep in mind that column loading capacities are not unlimited. The microgram to milligram range can be accommodated on the 2.1-mm i.d. columns without problems.

4.3 Peptide mapping

The determination of a total primary sequence by direct chemical means still involves the generation and isolation of numerous fragments of the parent molecule. Limited sequencing information for cloning purposes can be quickly generated from peptides. Relatively small fragments of 5–20 residues are easily isolated after proteolysis and

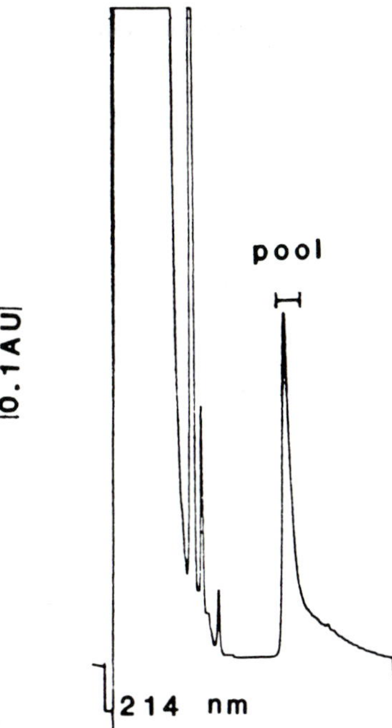

Figure 5. Sample de-salting by reverse-phase chromatography. Human cystatin S (2 μg) was alkylated using 4-vinylpyridine according to the given procedure (see text) and the resulting 52.5 μl injected onto a reverse-phase support (Aquapore™ RP-300, 2.1 × 30 mm). Elution was performed using a TFA/AN gradient. The peak (120 μl) was collected in an Eppendorf tube and recovered by lyophilization.

can be quickly sequenced. Larger fragments of 20–100 residues derived from chemical cleavages (CNBr at Met residues, acid cleavage at Asp-Pro, etc.) or limited enzymatic cleavages (at Lys, Arg, or other residues), are frequently more difficult to isolate in a homogeneous form. The reduced solubility of these larger denatured fragments introduces various isolation problems. Adding either guanidine hydrochloride or urea to the sample prior to injection appears to improve recoveries in some instances. There are even examples where such denaturants have been added to the chromatographic buffers. Adding such salts at high concentration to the buffers places certain demands on the HPLC equipment, for example requiring the use of corrosion-resistant instrumentation.

Generating peptide fragments from microgram or smaller amounts of protein requires special methodology. As with recovering samples from organic precipitation (see Section 3.8.1 above), it is necessary to maximize solubility. Simultaneously, large volumes must be avoided when the appropriate concentrations are needed for the digestion. *Table 11* indicates the buffers, denaturants and digestion conditions that function well on microgram amounts of material. The addition of a denaturant should be made from a concentrated solution, 6–8 M for guanidine hydrochloride or urea, 10% for SDS, which improves the likelihood of solubilizing the substrate. It might also be necessary to heat at 60–100°C to improve dissolution.

Table 11. Optimized protocols for micro-level enzymatic digestions.

Enzyme	*Digestion buffers*
Trypsin	0.1–0.25 M NH_4HCO_3, pH 7.8–8.1 ± 1–2 M urea
Chymotrypsin	As for trypsin
Staphylococcus aureus V8 protease	0.1–0.25 M NH_4HCO_3 or Tris acetate, pH 7.8–8.1 ± 0.01–0.1% SDS
Lys-C protease	As for V8 protease

Digestion conditions

1. Dissolve the samples in 25–50 μl for less than 10 μg, add the denaturant when solubility is a problem.
2. Add protease in 0.1–4% (w/w) ratio and digest at 25–50°C.
3. Perform narrow or micro-bore HPLC to follow digestion.
4. Preparative chromatography should be completed either directly on the digestion mixture or after stopping by a pH shift.

Protease quality is a major concern for extended digestions of more than a few minutes, incubations with high protease-to-substrate ratios or when extremely high cleavage specificity is needed (26). As an example, the protease trypsin is most frequently employed because of its high specificity for the peptide bond C-terminal to Arg and Lys, except when followed by Pro or, to more limited extents, when followed by an acidic residue like Asp, Glu or cysteic acid. Unfortunately the commercially available trypsin preparations, even those treated with inhibitors such as TPCK, are significantly contaminated with other proteases, in particular those with 'chymotryptic-like' activity.

Proteases frequently should be further purified before use. *Figure 6* illustrates the results one can expect from unpurified, commercially available TPCK-trypsin (*Figure 6B*) versus chromatographically purified material (*Figure 6A*). The pyridylethylated B-chain of insulin was digested for the indicated times and then directly chromatographed. Peaks numbered 1 and 3 are the exclusive products when digestion was limited to Arg^{22} and Lys^{29}; Ala^{30} elutes with the column breakthrough. Peaks designated a–e result from partial digestions of the N-terminal fragment (peak 3, residues 1–22), trypsin (peak 5) or both. The pattern illustrates the decreased peak heights for both the second major component and trypsin. In other words, overall recoveries are a direct function of contaminating protease activities.

4.4 **Selective methods for amino acid detection**

For certain purposes, particularly cloning, identifying fragments that contain specific amino acids is advantageous. The fragments most frequently sought contain Trp, Met and Cys, each of which can be detected using either direct or indirect procedures (*Table 12*). The easiest detection method is first UV scanning of the peaks as they elute, followed by an evaluation of the resulting spectra (9,10,27). Those peptides containing single, or sometimes multiple, aromatic residues (Trp, Tyr) or certain Cys derivatives (pyridylethyl-Cys) are identifiable. The major drawback to this method is the limited sensitivity and stability performances of photodiode detectors currently available. Most detectors are quite useful at the 100 pmol level but have severe performance problems at levels of 10 pmol or less.

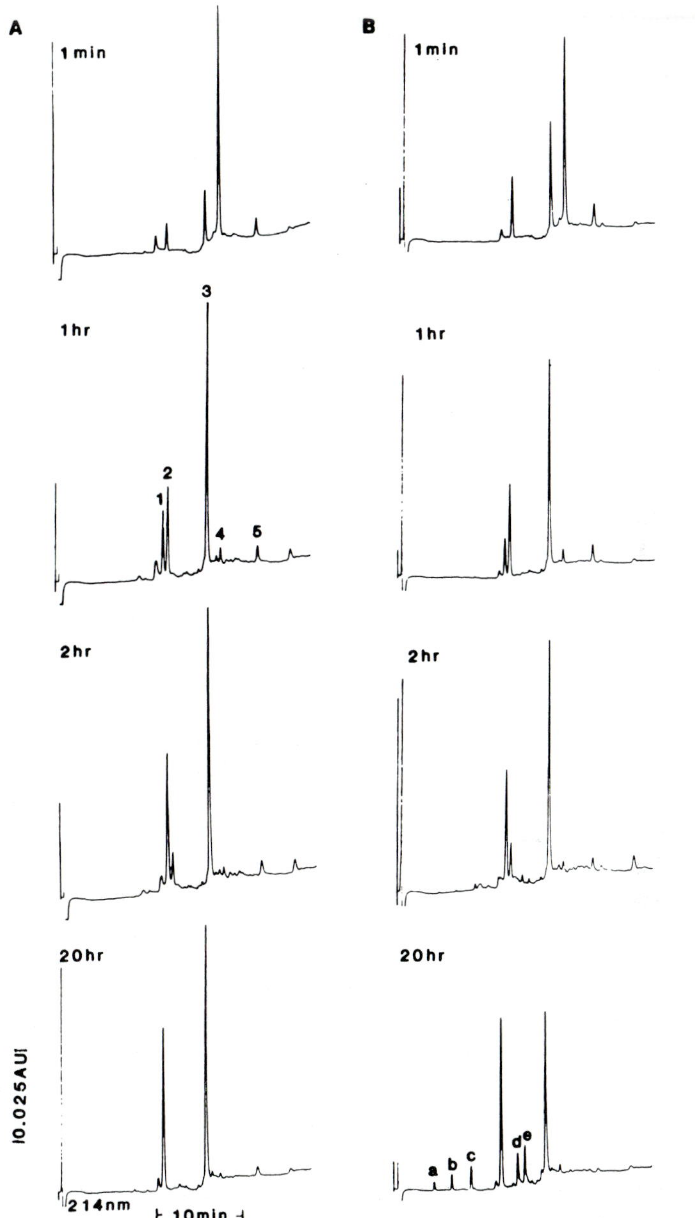

Figure 6. Peptide mapping of enzymatically derived products by HPLC. Bovine insulin was pyridylethylated according to the procedure given in the text. The products were isolated by reverse-phase HPLC and recovered by lyophilization (SpeedVac^{TM}). Because the alkylated B-chain is poorly soluble in the digestion buffer, the fragment was initially dissolved in 0.1% TFA and then a 1 μl aliquot (2 μg) diluted into 50 μl of 0.2 M ammonium bicarbonate, pH 8.1. Trypsin (0.2 mg/ml in 0.1% TFA) was added to give a 1:10 (w/w) enzyme-to-substrate ratio and aliquots of 10 μl (0.4 μg) injected at specified times. Digestions were carried out with HPLC-purified (**A**) or non-purified, commercially available (**B**) TPCK-trypsin. Peak identifications are: 1, residues 23–29; 2, residues 23–30; 3, residues 1–22; 4, intact PE-B-chain; 5, TPCK-trypsin. The identities of peaks a–e are unknown.

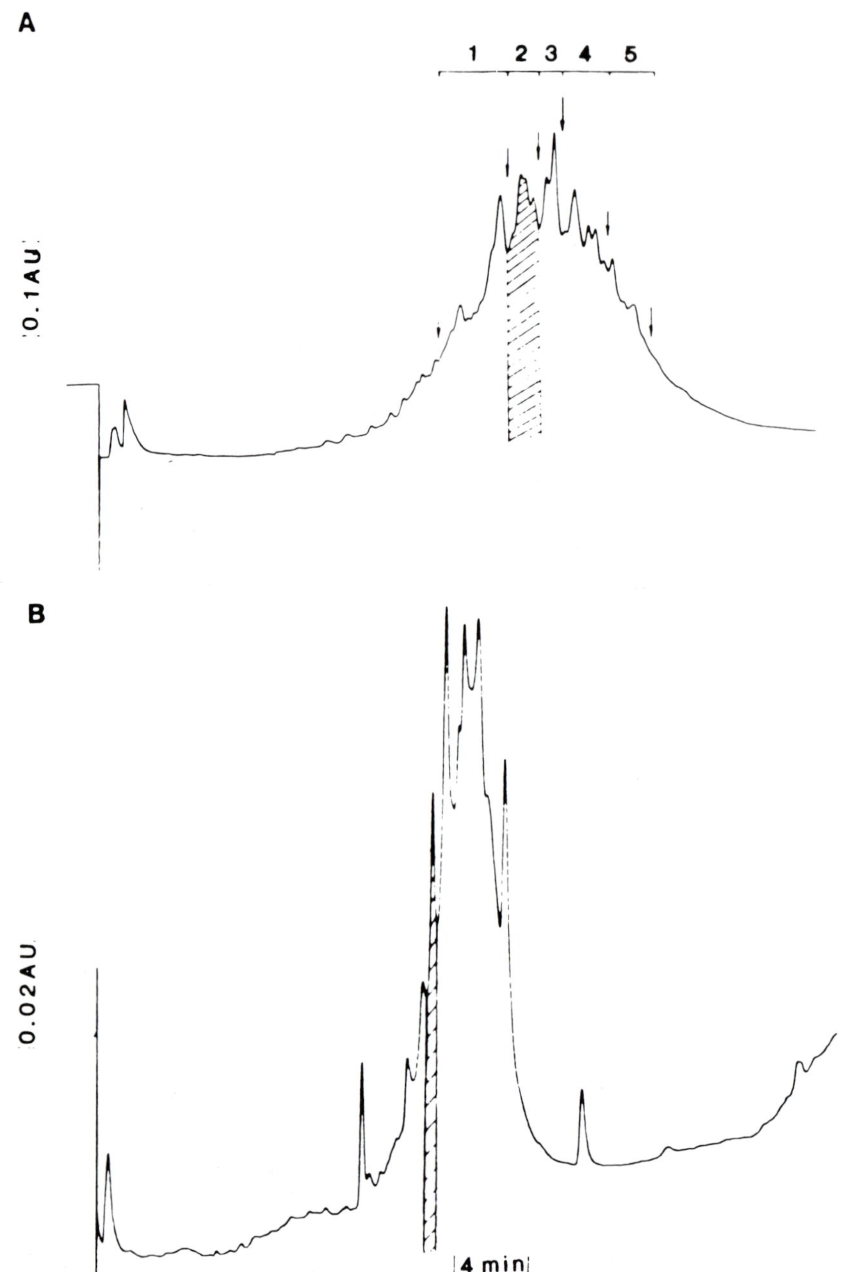

Figure 7. Selectivity modulation using different buffers. An atrial extract from pig heart was fractionated by size-exclusion chromatography and the below 5 kd range lyophilized. The resulting material was dissolved in 0.1% TFA and injected onto a reverse-phase support (Aquapore™ RP-300, 2.1 × 30 mm). (**A**) A linear gradient at 150 μl/min from 15% to 60% buffer B (A, 0.085% TFA; B, 70% AN in A) over 45 min was used to elute the indicated pools. (**B**) Following recovery by lyophilization, pool 2 was further fractionated on the same support using another buffer system (A, 15 mM ammonium acetate, pH 6.5; B, 70% AN in A). Elution was performed using a linear gradient at 150 μl/min from 15% to 45% B over 30 min.

Table 12. Selective identification methods for peptides or proteins containing specific amino acids.

	Method	*Residue*
Direct	UV scanning (photodiode)	Tyr, Trp, Cys
	Electrochemical	disulphides, Tyr
Indirect	Alkaline β-elimination	Phosphoserine
	Decarboxylation	α-carboxyGlu
	Ethoxyformylation	His
	Formylation	Tyr
	Iodination	Tyr
	S-methylation	Met
	O-nitrophenylsulphenylation	Trp
	Oxidation	Met, Trp
	Reduction	Disulphides

If relatively high detection sensitivities are required, one must perform analytical mapping prior to a preparative isolation. When available quantities are less than 100 pmol, it is difficult to work out optimized digestion conditions (enzyme−substrate ratio, times, etc.). In some cases narrow bore and micro-bore columns, and an optimized chromatographic system is essential.

4.5 **Selectivity modulation through buffer changes**

Frequently, even after the chromatographic separation is completed, the sample is still heterogeneous and alternative purifications are necessary. A common initial reaction is to change to another chromatographic mode and re-attempt the separation. There are, however, numerous other variables to consider (see *Tables 4* and *5*) before a column change is required.

With reverse-phase supports one such chromatographic change is simply to alter the pH of the elution buffer. Because the proteins and peptides chromatograph as charged species, their ionized states can be easily modulated. If the expected separation was not achieved at low pH (2−3), titrating to a higher value (pH 6−8) and re-injecting often helps.

Combining two different pH protocols is also possible. *Figure 7* illustrates how this adjustment was used to purify atrial peptides from pig heart extracts. An initial chromatographic separation was done using a TFA buffer (*Figure 7A*) followed by a separation using NH_4HCO_3 (*Figure 7B*). All other variables such as the column, running temperature and flow-rates were maintained constant. These particular buffers are quite useful because of their volatility under vacuum.

The sequence of the isolated peptide was easily determined (*Figure 8*). The elevated background of most PTH-amino acid derivatives in the first degradation cycle and its subsequent reduction in the second cycle, indicates the degree of contamination arising from the buffers. The degradative yield of 15 pmol Ala^2 also gives a 'feel' for the sensitivity levels possible with on-line PTH-amino acid identification. Such on-line analysis improves both the yields of certain labile derivatives and cycle to cycle chromatographic reproducibility.

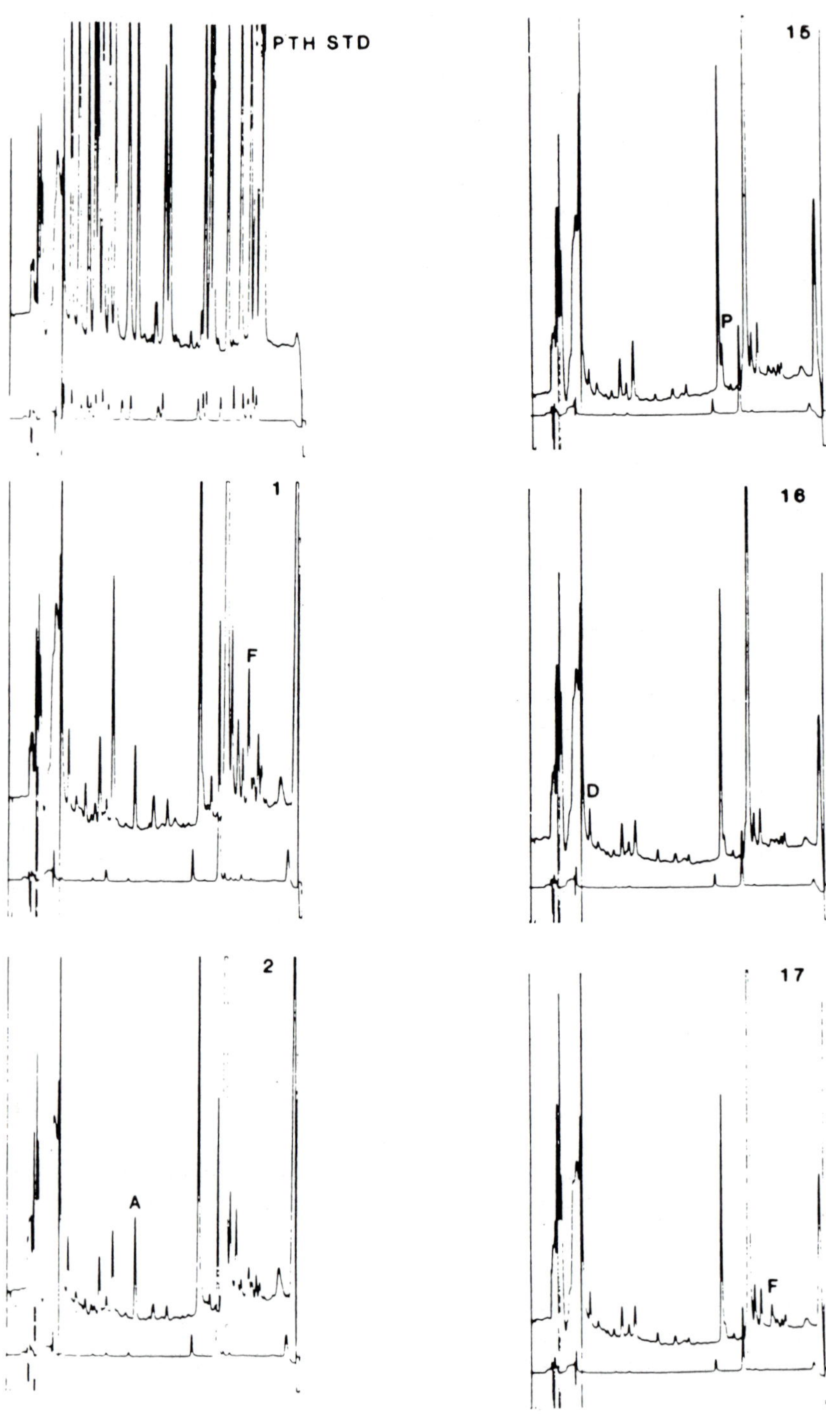
PTH STD
15
P
1
F
16
D
2
A
17
F

The buffer choice for the last chromatographic step in an isolation is important for subsequent characterization steps, an example might be the ammonia originating from NH_4HCO_3 and retained by samples recovered by lyophilization. This excess will often interfere with acceptable ammonia levels during amino acid analysis. In such cases, the TFA system is a better buffer choice.

The salts remaining from other separation modes present similar problems and can interfere with subsequent steps. When isolations are carried out on ion-exchange or hydrophobic interaction supports, different salts can be tested while other conditions are maintained constant. Numerous examples are given in the literature that illustrate this form of solute modulation (28–33).

4.6 Two-dimensional chromatographic isolation of albumin

Combining an initial ion-exchange separation with a reverse-phase step has been recognized as a practical arrangement for an isolation scheme. The two modes separate on the basis firstly of difference in charge and secondly in hydrophobicity. Arranging chromatography in this order provides a number of advantages: quick sample de-salting, reduced losses and sample concentration. By carrying out the final elution in a volatile buffer, one easily recovers the material in a 'condition' ready for further characterization by sequencing, gel electrophoresis or amino acid analysis.

Figure 9 illustrates the cation-exchange separation of 250 μg of rabbit muscle crude extract (34). The weighed sample was readily soluble in the chromatographic buffer at reasonably high concentrations (5 mg/ml). Two additional 250 μg preparations were performed and the first peaks (shaded area in *Figure 9A*) combined from each. The pools were lyophilized and re-dissolved in 0.1% TFA (100 μl).

An analytical reverse-phase run on 5 μl aliquots was performed to optimize chromatography and then the remaining material preparatively chromatographed (*Figure 9B*). Following sample recovery by drying, the pools were characterized by SDS–PAGE and sequencing. Section 5.3 will compare these results with those from a two-dimensional chromatographic/electrophoretic isolation of the same material.

5. ELECTROPHORESIS APPLICATION EXAMPLES

5.1 Expected recovery yields

The SDS–PAGE electroblotting procedures (21,23) provided a unique way to transblot proteins onto the appropriate surfaces. In sequence analyses of the blotted proteins we have noted that the non-covalent immobilization is generally insufficient to prevent extractive losses. This results in lower repetitive yields for most of the samples analysed (19,34).

Figure 8. Sequence determination of peptide 4 from atrial extract. The entire lyophilized sample was sequenced in a gas-phase instrument using an on-line PTH-analyser for derivative identification. A PTH-amino acid standard (25 pmol) is illustrated (upper left), as well as the first 2 and final 3 cycles from the degradation. Detection was carried out at 269 nm; 40% of the sample from each cycle was chromatographed. The sequence was: Phe-Ala-Val-Asp-Tyr-Ser-Lys-Leu-Lys-Lys-Glu-Gly-Pro-Asp-Phe.

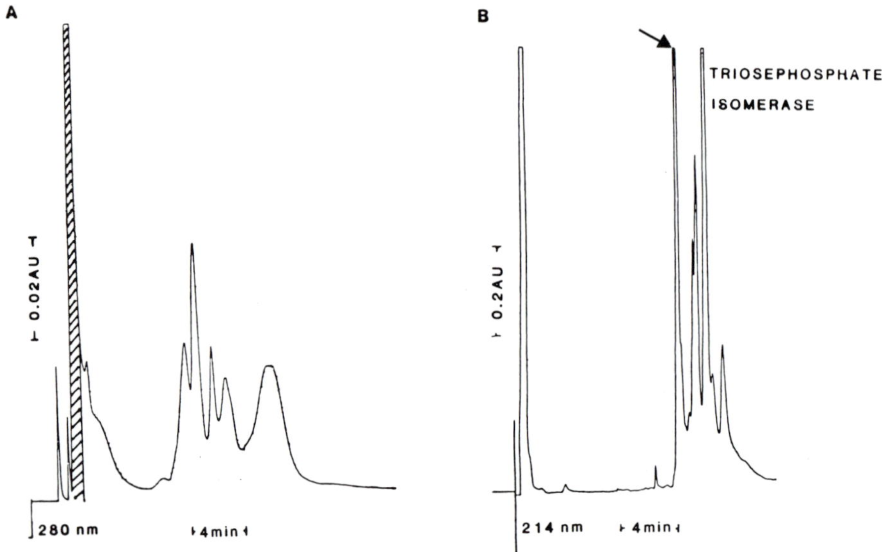

Figure 9. Two-dimensional chromatographic isolation of albumin from crude rabbit muscle extract (34). (**A**) 250 μg of extract was dissolved in 50 μl of 20 mM sodium phosphate, pH 6.0 and injected into a cation exchanger (Aquapore™ CX-300, 2.1 × 30 mm) equilibrated in the same buffer. Elution was performed isocratically at 0% B for 5 min prior to introducing a 20 min linear gradient to 100% B (0.5 M NaCl in buffer A). Flow-rate was 200 μl/min at room temperature. The indicated peak (shaded area) was collected for further purification. (**B**) The combination of three isolations, 750 μg, were pooled, lyophilized and re-dissolved in 100 μl of 0.1% TFA. The majority of sample (95 μl) was chromatographed on a reverse-phase support (Aquapore™ RP-300, 2.1 × 30 mm) using a 30 min linear gradient from 0.1% TFA (buffer A) to 70% AN in 0.1% TFA (buffer B). The designated peak (albumin, arrow) was collected and sequenced (see *Figure 15* and *Table 15* for results).

5.1.1 *Derivatized glass surfaces*

The key modifications to the Aebersold *et al.* (22) procedures for derivatized glass surfaces, described in Section 3 (above), are:

(i) replacing acetone with acetonitrile in the preparation of the derivatized filters;
(ii) using *N*-ethylmorpholine instead of Tris-glycine as the electroblotting buffer; and
(iii) sequencing the blotted samples in the presence of polybrene.

In our hands (19,34) these modifications significantly improve repetitive yields at low microgram levels.

Figure 10 indicates the recoveries of various amounts of β-lactoglobulin A, electroblotted using the described protocol. Also illustrated are two additional recoveries that represent 100% theoretical efficiency, that is an absolute 100% sequencing yield and 100% recovery from SDS–PAGE electroblotting. Efficiency of 50% can be explained by a 50% reduction (loss) of either the sequencing absolute yield or SDS–PAGE electroblotting. The theoretical plot assumes quantitative recoveries from all steps: sample handling, electrophoresis, electroblotting and sequencing. In actuality, one normally experiences recoveries in the 50% range from the sequencing step itself. The difference between the approximately 50% sequencing efficiency and the

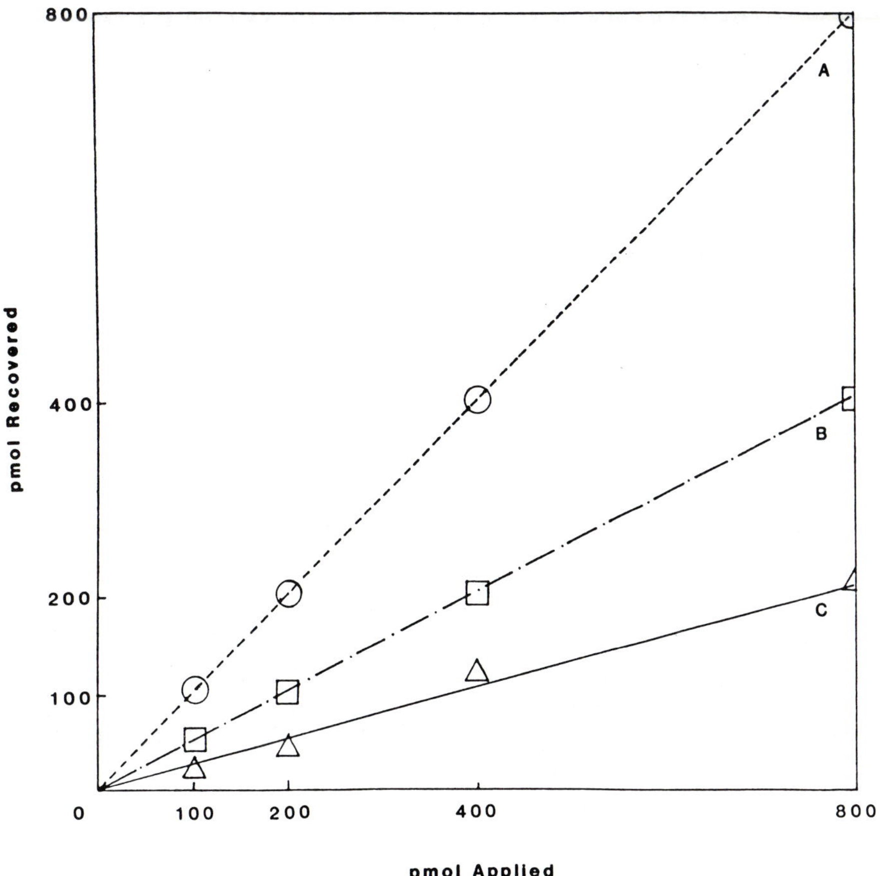

Figure 10. Sequencing recoveries from electrophoresed β-lactoglobulin A (19). Different amounts of the protein were separated by SDS−PAGE and electroblotted onto AP GF-F glass fibre filters. The bands were excised from the primary sheet and recoveries determined by their initial yields. Assuming no losses of totally sequenceable material, **line A** represents 100% recovery efficiency. **Line B** depicts the usual 50% efficiencies found with 'typical' samples which have not been isolated by SDS−PAGE electroblotting. **Line C** shows the current results.

β-lactoglobulin recoveries, approximately 25% of total, represent losses from sample handling and/or amino-terminal blockage.

Another way of presenting this data is shown in *Figure 11*. Here, recoveries onto both AP GF-F and QA-F surfaces are compared with the sequenceable amounts, 50% of total applied. The average capacity of the derivatized sheets was approximately 20−25 μg/cm^2. Typically, protein bands up to 15 μg are found primarily on the first sheet of derivatized filter.

The optimum electroblotting time was 30−60 min at room temperature for β-lactoglobulin A (*Figure 12*). Proteins with molecular weights up to 90 kd can be transblotted within 60 min. Under prolonged electroblotting of 2−4 h, samples slowly

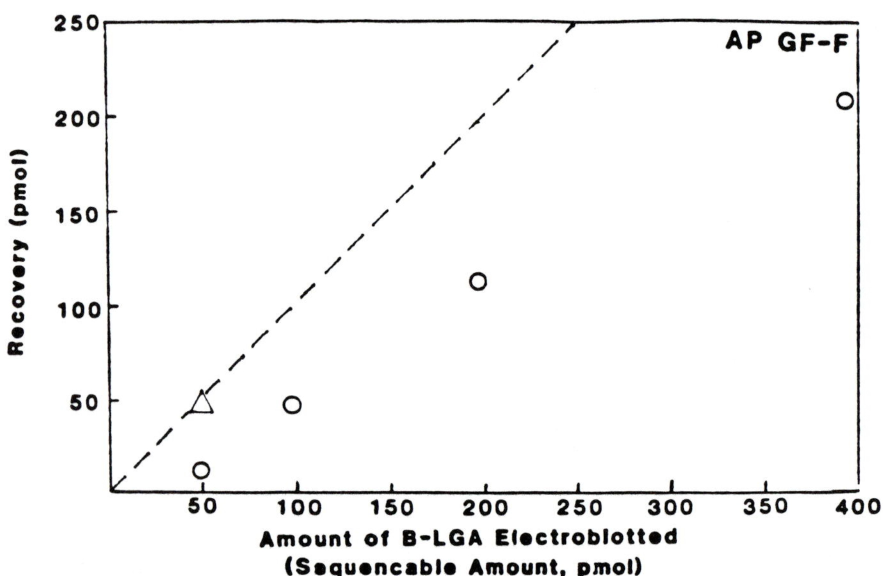

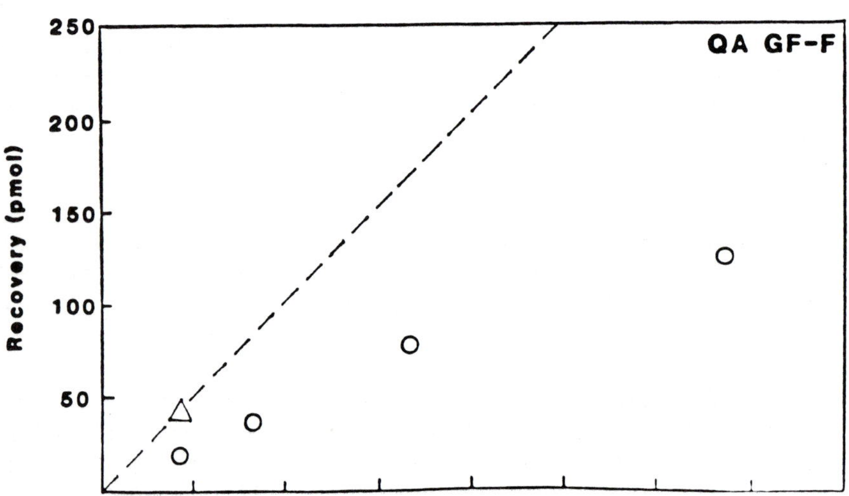

Figure 11. Recovery versus amount electrophoresed and electroblotted (19). Different amounts of β-lactoglobulin A were separated by SDS–PAGE and electroblotted onto QA GF-F or AP GF-F glass filter sheets. The protein bands found in the first sheet were excised, sequenced, and recoveries determined by initial sequencing yields. The dotted lines represent the theoretical 100% recovery up to 250 pmol loadings. A single experimental point (△) was determined.

penetrate through to the secondary sheet. Sample losses are possible from diffusion either out of the gel, or derivatized filter or through the filter during prolonged blotting. One should therefore attempt to minimize blotting times. Evidently 'over-transferring' can occur and possibly cause losses between the gel and blotting surfaces.

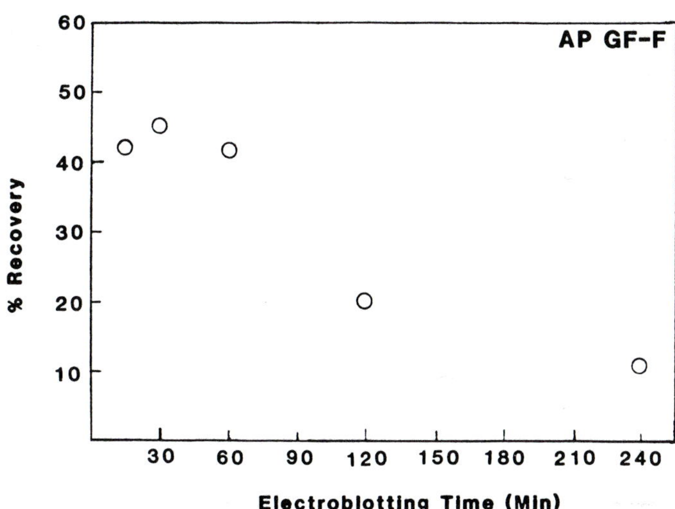

Figure 12. Electroblotting time versus recovery (19). 200 pmol of β-lactoglobulin A was separated by SDS–PAGE and electroblotted for the indicated times. The bands found on the first sheet were excised and sequenced. The % recovery calculations were based on the initial yields of sequence analyses before and after isolation.

Figure 13 shows the effects of polybrene on repetitive yields of electroblotted β-lactoglobulin A. Increases of 3–4% in repetitive yields were found for both quaternary ammonium and aminopropyl-derivatized glass filters. The addition of polybrene traps blotted sample that would otherwise be lost during the solvent extraction steps of sequencing.

5.1.2 *Polyvinylidene difluoride (PVDF)*

The results summarized in *Table 13* indicate that sequencing yields, and therefore mass yields, from PVDF membranes are marginally higher than from either the AP GF-F or Polybase glass supports. Initial sequencing yields indicate that approximately 50–60% of the sample was recovered, as compared with the control, and that the percentage appeared to be independent of loading amounts.

Sequencing efficiencies, or more specifically repetitive yields, were evaluated by maintaining sample amounts constant (200 pmol) and electroblotting onto the various membranes for direct sequence analysis. As summarized in *Table 14*, the PVDF membrane was the only support with an acceptable repetitive yield, provided the sequencing was performed on the Model 470A Gas Phase Sequencer.

Insertion of a polybrene-pre-conditioned filter beneath the blotted pieces, improved the repetitive yields in either sequencer by 4–10% and indicated that the low efficiencies were due to sample washout. The ionic interactions between the polypeptides and the chemically derivatized glass fibres apparently fail to provide sufficient affinity to prevent solvent extractive losses. On the other hand, the hydrophobic binding of protein onto PVDF proves that attraction. Samples can, however, be dislodged during the acid cleavage step of the Pulse Liquid Phase Sequencer. This is probably due to the lower

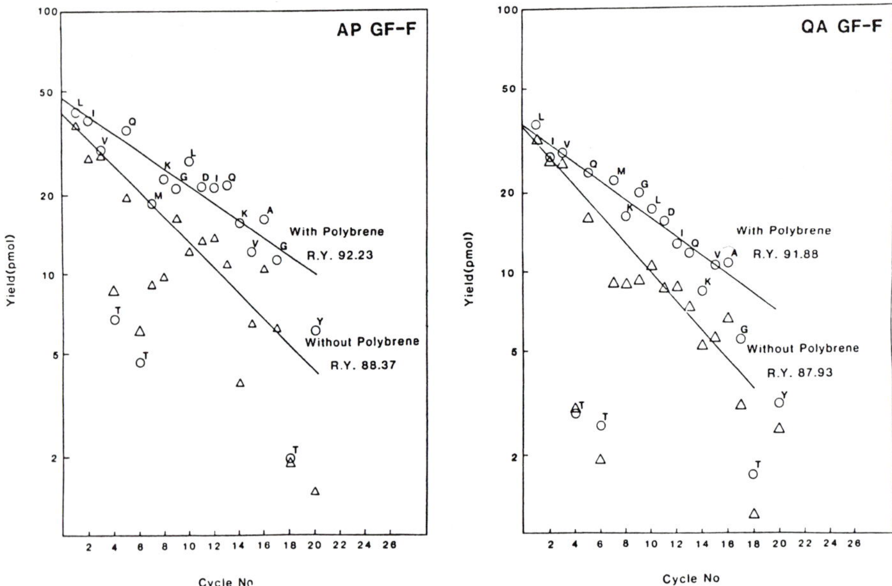

Figure 13. Sequence analyses of electroblotted β-lactoglobulin A with and without polybrene (19). 200 pmol of β-lactoglobulin A was separated by PAGE and electroblotted onto a QA GF-F or AP GF-F filter. Sequence analysis was carried out in the presence (○) or absence (△) of polybrene.

Table 13. Sequencing recoveries from different blotting membranes.

Sample applied (pmol)	*Recovery (%)*[a]			
	Control	*AP GF-F*	*Polybase*	*PVDF*
100	57.9 ± 7.0 (n = 8)	24.8 ± 11.8 (n = 4)	22.4 ± 1.3 (n = 4)	31.6 ± 5.4 (n = 4)
200	56.9 ± 15.6 (n = 2)	26.9 ± 9.4 (n = 8)	32.6 ± 7.4 (n = 4)	34.3 ± 8.5 (n = 11)
400		30.8 ± 13.3 (n = 2)	27.7 ± 9.5 (n = 2)	32.7 ± 9.1 (n = 3)
800		26.9 ± 11.5 (n = 3)	33.3 ± 2.8 (n = 2)	34.6 ± 1.8 (n = 2)

[a]Calculated from the initial yields of PTH-Leu.

volume capacity of PVDF, versus GF-C glass filter, for the acid.

In a forthcoming User Bulletin (35) we review the results of a series of experiments that compare the blotting capacities and sequencing efficiencies of three different blotting membranes—aminopropyl glass fibre sheet, polybase-coated glass fibre sheet, and Immobilon membrane. To broaden the application scope of the PAGE–electroblotting technique in protein structural analysis, experimental protocols for on-membrane amino acid analysis, protein extraction from membrane and *in situ* chemical and proteolytic cleavages are also included.

Table 14. Sequencing efficiencies of electroblotting membranes.

Membrane/sequencer	*Direct*	*Insertion of a polybrene-conditioned filter*
AP GF-F/470A	88.3	92.2
Polybase/477A	79.8	88.9
PVDF/470A	95.5	ND
PVDF/477A	83.1	95.1

ND, not determined.

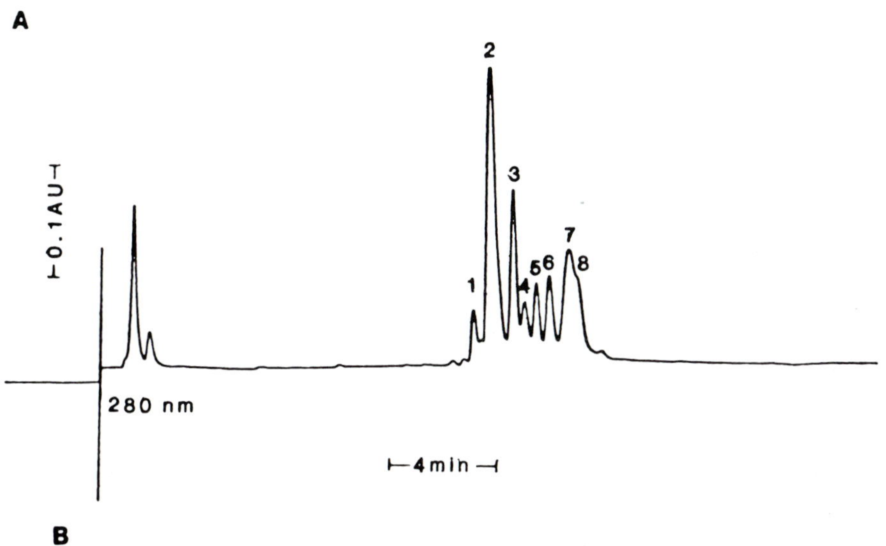

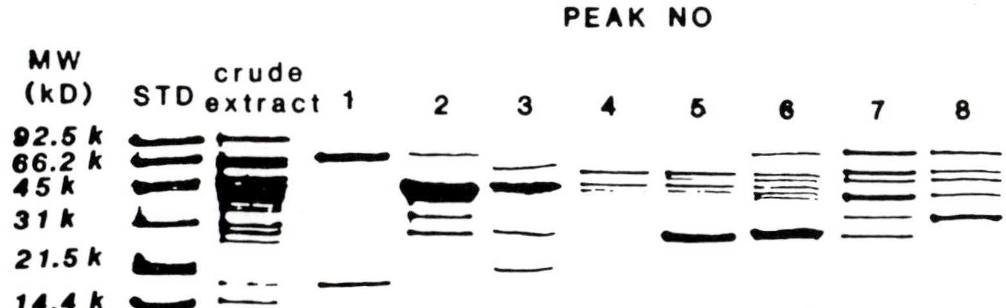

Figure 14. Two-dimensional chromatographic–electrophoretic isolation of albumin from crude rabbit muscle extract (34). **(A)** 250 μg of extract was dissolved in 50 μl of 0.1% TFA, injected onto a reverse-phase support (Aquapore[TM] RP-300, 2.1 × 30 mm) and eluted as described in *Figure 10B*. The individual peaks were collected, lyophilized and re-dissolved in gel sample buffer. **(B)** Recovered fractions from the first step (250 μg) were further separated on an SDS mini-gel (0.5 × 60 × 90 mm, a 5% stacking and 15% resolving Laemmli system) and stained with Coomassie blue. The same amount of isolate (250 μg) was similarly fractionated, electroblotted and located using the hydrophobic fluorescent dye.

5.2 Two-dimensional chromatographic/electrophoretic isolation of albumin

Combining both chromatographic and electrophoretic separation modes into an isolation scheme offers a number of advantages. Perhaps the most important is the ease of sample recovery from the chromatographic step, especially when a reverse-phase isolation is

Table 15. Sequencing yields of albumin isolated from rabbit muscle extract by either two-dimensional chromatography or a chromatography–electrophoresis combination[a].

Cycle no.	*Amino acid*	*Yields (pmol) from*	
		IEC/RPC[b,c]	*RPC/SDS–PAGE*[b]
1	E	17.8	32.2
2	A	24.8	25.2
3	H	a	6.9
4	K	6.5	22.0
5	S	2.9	4.4
6	E	5.2	18.2
7	I	11.5	15.3
8	A	8.5	14.5
9	H	a	3.7
10	R	a	5.8
11	F	9.6	8.2
12	N	2.8	9.1
13	D	1.8	5.2
14	V	3.8	5.2
15	G	3.7	3.1
16	E	1.9	5.2
17	E	a	7.5
18	H	a	0.7
19	F	2.8	2.6
20	I	4.6	2.7
21	G	1.2	2.3
22	L	5.1	2.2
23	V	3.0	1.8
24	L	2.3	2.7
25	I	3.0	1.9
26	T	2.1	
27	F	4.6	
28	S	0.8	
29	Q	0.9	
30	Y	0.8	
31	L	2.7	
32	Q	0.6	

[a]Data taken from refs 19 and 34.
[b]Order of separation steps was: IEC followed by RPC and RPC followed by SDS–PAGE plus electroblotting.
[c]An 'a' indicates its presence but not calculated.

peformed. Because electrophoresis has a molecular size limit below which smaller proteins and peptides are difficult to separate, one can reasonably precede such a step with HPLC which performs well on the smaller compounds.

The reverse-phase analytical separation of 250 μg of crude rabbit muscle extract is illustrated in *Figure 14*. The degree of fractionation is quite similar to that found for the cation-exchange separation of the same material (*Figure 9*). There were 6–7 well-defined peaks in both cases. SDS–PAGE analysis of all the material from each fraction indicated a high degree of heterogeneity in most pools except RP-2, RP-4 and RP-5, where only minor contaminants are noted (*Figure 14B*).

Following another reverse-phase isolation which started with 250 μg sample, each fraction was subjected to SDS–PAGE and the bands electroblotted onto an AP-glass filter. Following location using the hydrophobic stain, the 66-kd band from fraction 1 was excised and sequenced (see following Section 5.3).

5.3 **Amino-terminal sequencing yields**

Table 15 summarizes the residue identifications and sequencing yields from the degradations of albumin isolated by the two different two-dimensional procedures. Additionally, the PTH-analyses of cycles 1–3 and 19–21 from each sample are illustrated in *Figure 15*. Non-normalized initial yields were 17 pmol and 32 pmol from the two-dimensional chromatography versus the chromatography/electrophoresis combination, respectively.

Background contaminations in the first sequencing cycle were essentially the same for both two-dimensional methods. One can first normalize the results in *Figure 15* by doubling the peak sizes in those chromatograms from the CX/RP isolation. It is then clear that initial contaminations were similar in both. Although not shown for any of the sequencing runs, the repetitive yields for all samples (see *Table 16*) was approximately 90% for the samples exhibiting initial yields ranging from 6 to 80 pmol. The isolation method did not seem to influence significantly the repetitive yield of any of the sequenced proteins.

Sequencing was performed on four additional proteins also isolated at the same time as the albumin. Significant differences in initial yields were observed for some of the isolates (*Table 16*). The total recovery of three proteins, triosephosphate isomerase, albumin and phosphoglucomutase, were similar and higher recoveries were found for aldolase by the CX/RP modes and for creatine kinase by the RP/SDS–PAGE combination. With the yields of three of the five proteins approximately equal, it appears that differential amino-terminal blockage was not occurring in these cases. With the two other proteins, however, the recovery differences might be explained by the yield from the individual isolation steps: CX, RP and/or SDS–PAGE. The amino-terminal sequences of the proteins agreed with previously reported sequences or, in the case of creatine kinase, mRNA sequence translation. Of the 32 amino-terminal residues identified in the muscle albumin, only eight differed from the corresponding region of human albumin.

Besides addressing the overall yields, or 'sequenceabilities', these experiments also provide insight into the resolution limitations of the two different isolation schemes. The initial fractionations by either cation-exchange (*Figure 9A*) or reverse-phase HPLC (*Figure 14A*) separated approximately the same number of peaks. The additional reverse-phase separation (*Figure 10B*) of the CX pools resulted in the isolation of approximately 20 peaks. A similar number of bands was detected by SDS–PAGE (*Figure 14B*) of the fractions first separated by RP-HPLC. In other words, a comparable number of components was observed for both isolation procedures.

6. SUMMARY

The combination either of microbore liquid chromatography with gel electrophoresis and electroblotting, or of two different micro-chromatographic separations is extremely effective in preparing isolates in the microgram range. Clearly, the use of either can be equally effective, provided homogeneity is attained in a single step.

A flow chart depicting when and where the liquid chromatography and/or electrophoretic techniques should be used is illustrated in *Figure 16*. Assuming that

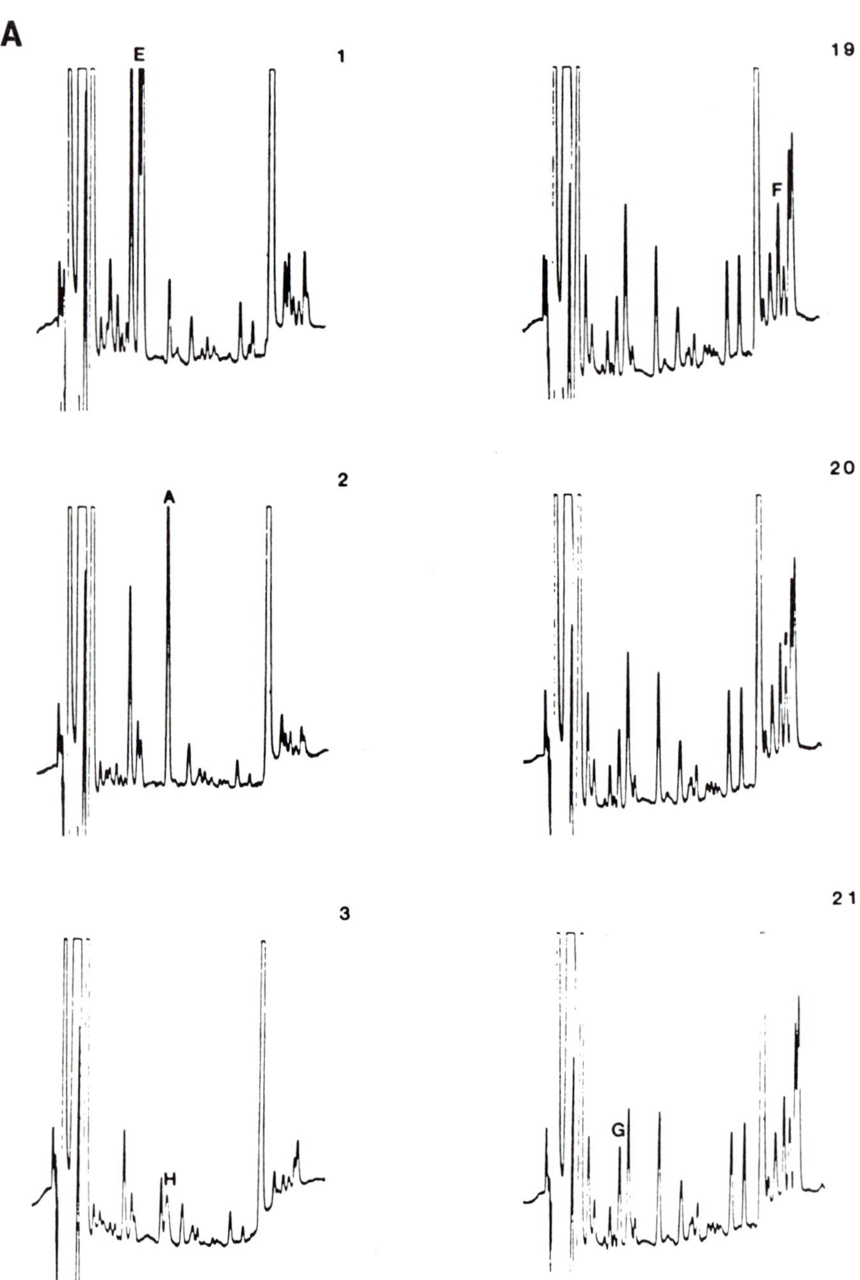

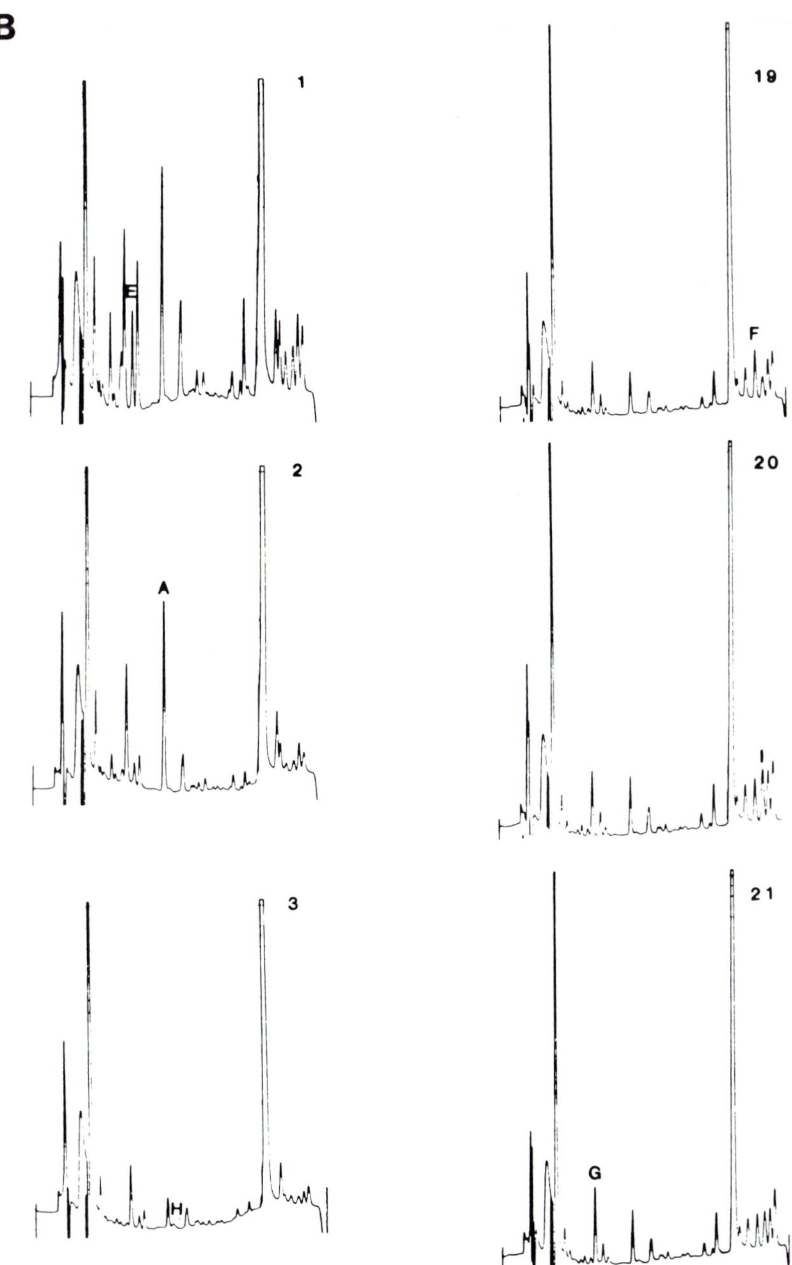

Figure 15. Amino-terminal sequences of rabbit muscle albumin isolated by different two-dimensional methods (19,34). Samples were fractionated by IEC/RPC (**A**) or RPC/electrophoresis (**B**). Column effluents were collected into Eppendorf tubes and recovered by lyophilization. Electroblotting was used to elute samples following electrophoretic separation. PTH-amino acids were analysed on-line by chromatographing 40% of the product of each cycle on a reverse-phase support. The first three and last three cycles of each preparation are illustrated.

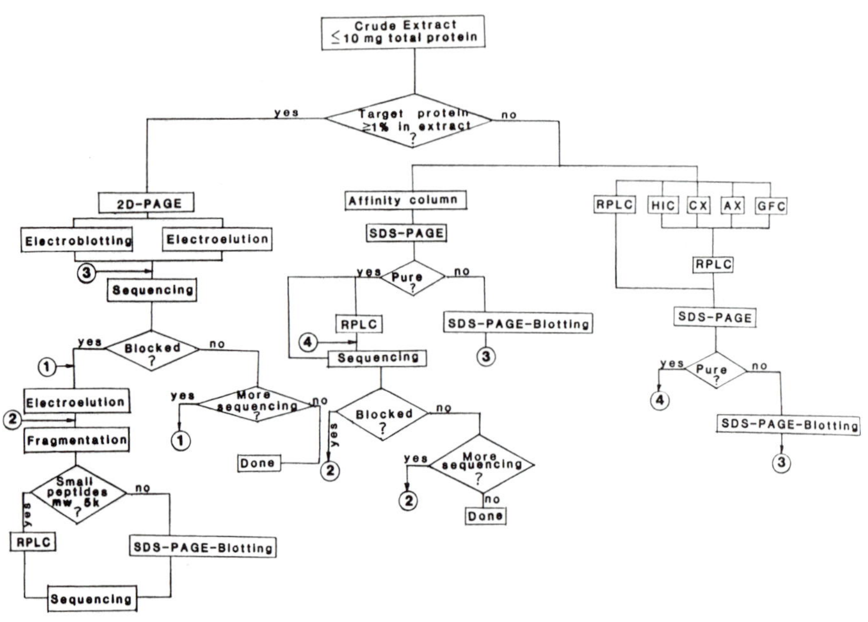

Figure 16. Flow diagram of electrophoretic/chromatographic applications in sample isolation.

Table 16. Sequence analysis of proteins isolated by two-dimensional methods[a].

Protein	*Total initial yield (pmol)*[b]	
	IEC/RPC	*RPC/SDS–PAGE*
Triosephosphate isomerase	198	176
Aldolase	185	61
Creatine kinase	58	234
Albumin	61	97
Phosphoglucomutase	22	21

[a]Data derived from ref. 34.
[b]Total initial yield (or recovery) calculated from N-terminal sequencing of each isolated sample.

only 10 mg of starting sample is available, one can conveniently use either method or their combination to isolate microgram quantities successfully. There are necessary precautions to be addressed prior to an effective implementation of the methodology, but once resolved micro-level characterizations can be effectively carried out.

7. ACKNOWLEDGEMENTS

The authors would like to take this opportunity to thank various collaborators, in particular Sylvia Yuen, for their help during the investigations covered in this chapter.

8. REFERENCES

8.1 References of general interest for protein/peptide isolation by HPLC

Hughes,G.J. and Wilson,K.J. (1983) High-performance liquid chromatography: analytic and preparative applications in protein–structure determination. In *Methods in Biochemical Analysis.* Glick,D. (ed.), John Wiley and Sons, Inc., New York, Vol. 29, p. 59.

Hunkapiller,M.W., Strickler,J.E. and Wilson,K.J. (1984) Contemporary methodology for protein structure determination. *Science,* **226**, 304.

Novotny,M. (1985) Microcolumn liquid chromatography. Biomedical and pharmaceutical applications. *LC, Liquid Chromatogr. HPLC Mag.*, **3**, 876.

Matson,R.S. and Goheen,S.C. (1987) HPLC separation of membrane proteins. *LC-GC, Mag. Liquid Gas Chromatogr.*, **4**, 624.

Wittliff,J.L. (1986) HPLC of steroid-hormone receptors. *LC-GC, Mag. Liquid Gas Chromatogr.*, **4**, 1092.

8.2 References in text

1. Schwartz,H.E. and Brownlee,R. (1984) *Am. Lab.*, **16**, 43.
2. Schwartz,H.E. and Berry,V.V. (1985) *LC, Liquid Chromatogr. HPLC Mag.*, **3**, 110.
3. Dolan,J.W. (1985) *LC, Liquid Chromatogr. HPLC Mag.*, **3**, 92.
4. Schwartz,H.E. (1986) *J. Chromatogr. Sci.*, **24**, 285.
5. Schwartz,H.E. and Brownlee,R.G. (1985) *J. Chromatogr. Sci.*, **23**, 402.
6. Wilson,K.J. (1987) In *Protein/Peptide Sequence Analysis: Current Methodologies.* Bhown,A. (ed.), CRC Press, Boca Raton, Florida, p.1.
7. Wilson,K,J., Yuan,P.M. and Schlabach,T.D. (1989) In *The Use of HPLC in Protein Purification and Characterization.* Kerlavage,A.R. (ed.), A.R.Liss, Inc., New York, in press.
8. Nice,E.C., Lloyd,C.J. and Burgess,A.W. (1984) *J. Chromatogr.*, **296**, 153.
9. Grego,B., Van Driel,I.R., Stearne,P.A., Goding,J.W., Nice,E.C. and Simpson,R.J. (1985) *Eur. J. Biochem.*, **148**, 485.
10. Nice,E.C., Grego,B. and Simpson,R.J. (1985) *Biochem. Int.*, **11**, 187.
11. Pearson,J.D. (1986) *Anal. Biochem.*, **152**, 189.
12. Wilson,K.J., Van Wieringen,E., Klauser,S. and Berchtold,M.W. (1982) *J. Chromatogr.*, **237**, 407.
13. Hughes,G.J. and Wilson,K.J. (1983) *Methods Biochem. Anal.*, **29**, 59.
14. Trumbore,C.N., Tremblay,R.D., Penrolse,J.T., Mercer,M. and Kelleher,F.M. (1983) *J. Chromatogr.*, **280**, 43.
15. Sadek,P.C., Carr,P.W., Bowers,L.D. and Haddad,L.C. (1985) *Anal. Biochem.*, **144**, 128.
16. Abbott,S.R. (1986) *LC-GC, Mag. Liquid Gas Chromatogr.*, **4**, 522.
17. Wilson,K.J., Hong,A.L., Brasseur,M.M. and Yuan,P.M. (1986) *Biochromatography,* **1**, 106.
18. Schlabach,T.D. and Wilson,K.J. (1987) *J. Chromatogr.*, **385**, 65.
19. Yuen,S., Hunkapiller,M.W., Wilson,K.J. and Yuan,P.M. (1986) User Bulletin No. 25, Applied Biosystems, Inc., Foster City, California.
20. Hunkapiller,M.W, Lujan,E., Ostander,F. and Hood,L.E. (1983) *Methods in Enzymology,* Hirs,C.H.W. and Timasheff,S.N. (eds), Academic Press, New York, Vol. 91, p. 227.
21. Vandekerckhove,J., Bauw,G., Paype,M., VanDamme,J. and Van Montague,M. (1985) *Eur. J. Biochem.*, **152**, 9.
22. Aebersold,R.H., Teplow,D.B., Hood,L.E. and Kent,S.B.H. (1986) *J. Biol. Chem.*, **261**, 4229.
23. Matsudaira,P. (1987) *J. Biol. Chem.*, **262**, 10035.
24. Hames,B.D. and Rickwood,D. (eds) (1981) *Gel Electrophoresis of Proteins: A Practical Approach.* IRL Press, Oxford and Washington, DC.
25. Isemura,S., Saitoh,E. and Sanada,K. (1984) *J. Biochem.*, **96**, 489.
26. Titani,K., Sasagawa,T., Resing,K. and Walsh,K.A. (1982) *Anal. Biochem.*, **123**, 408.
27. Yang,C.-Y., Pownall,H.J. and Gotto,A.M., Jr. (1985) *Anal. Biochem.*, **145**, 67.
28. Kopaciewicz,W. and Regnier,F.W. (1983) *Anal. Biochem.*, **133**, 151.
29. Gooding,K.M. and Schmuck,M.N. (1984) *J. Chromatogr.*, **296**, 321.
30. Melander,W.R., Corradini,D. and Horvath,C.S. (1984) *J. Chromatogr.*, **317**, 67.
31. Gooding,D.L., Schmuck,M.N. and Gooding,K.M. (1984) *J. Chromatogr.*, **296**, 107.
32. Fausnaugh,J.L., Kennedy,L.A. and Regnier,F.E. (1984) *J. Chromatogr.*, **317**, 141.
33. Brandts,P.M., Gelsema,W.J. and D.DeLigny,C.L. (1985) *J. Chromatogr.*, **333**, 41.
34. Yuen,S., Hunkapiller,M.W., Wilson,K.J. and Yuan,P.M. (1988) *Anal. Biochem.*, **168**, 5.
35. Yuen,S.W., Chui,A.H., Wilson,K.J. and Yuan,P.M. (1988) User Bulletin No. 35, Applied Biosystems Inc., Foster, City, CA.

CHAPTER 2

Peptide preparation and characterization

A.AITKEN, M.J.GEISOW, J.B.C.FINDLAY, C.HOLMES and A.YARWOOD

1. AMINO ACID COMPOSITION ANALYSIS

Amino acid composition data are of value in microsequencing only if reliable analyses can be obtained on amounts of material compatible with the sensitivity of current automated sequencers. Unless this is the case, material in short supply is better devoted to sequence analysis. Nevertheless, quantitation is needed where proteins/peptides are N-terminally blocked, where post-translational modifications of certain types are suspected or where prior screening of a large number of peptides may assist in identifying suitable candidates for sequence analysis.

Current popular single-column ion-exchange amino acid analysers can be operated down to a sensitivity limit of about 50-100 pmol using ninhydrin-based detection and below this using *o*-phthalaldehyde (OPA) or fluorescamine. Advantages of ion-exchange-based analysis lie in relative tolerance to impurities because of post-column derivatization chemistry. The very long development period over which the method has developed means that the process has been optimized and the elution behaviour of most physiological amino-containing compounds is known. The requirements of speed and sensitivity have recently led to the development of reverse-phase systems but these are not yet as robust as the more traditional ion-exchange methods.

1.1 **Ion-exchange systems**

1.1.1 *Sample preparation/hydrolysis*

(i) Add a known amount of a suitable internal standard (e.g. norleucine) to the sample in a thick-walled borosilicate glass tube and dry by rotary evaporation or centrifugation under reduced pressure.

(ii) Re-suspend the sample in 0.1 – 1.0 ml of 5.7 M HCl (constant boiling) containing 0.1% phenol.

(iii) Draw out the neck of the tube in an oxygen/gas flame to leave a capillary. Evacuate and flush the tube with O_2-free N_2 by means of a silicon rubber tube connected to an oil pump and trap and a regulated N_2 cylinder via a three-way stopcock. Bumping can be discouraged by freezing the sample.

(iv) When there is no further frothing, seal the tube at the constriction with a flame and hydrolyse the contents at 110°C for 18 – 96 h as required.

(v) After hydrolysis score the tubes with a diamond scriber, open by touching the score with a hot Pyrex rod and either rotary evaporate or dry *in vacuo* in a desiccator over sodium hydroxide.

Vapour-phase hydrolysis is less tedious when dealing with large numbers of samples

Table 1. Hydrolysis methods.

	Conditions	*References*
Rapid hydrolysis	HCl/propionic acid (50:50 v/v), 160°C, 15–30 min, *in vacuo*.	1
Amino acids		
Met, Cys, Tyr	6 M HCl/Na_2SO_4, 110°C, 24 h, *in vacuo*.	2
Trp	3 M *p*-toluenesulphonic acid + 0.2% tryptamine, 24 h, 110°C, *in vacuo*.	3
	or 3 M mercaptoethanesulphonic acid, 110°C, 24–72 h, *in vacuo*.	4
	or 4 M methanesulphonic acid + 0.2% 3-(2-aminoethyl) indole, 115°C, 24–72 h, *in vacuo*.	5
o-Phospho-Ser	6 M HCl, 110°C, 1–2 h.	
o-Phospho-Thr	6 M HCl, 110°C, 2–4 h.	6, 7
o-Phospho-Tyr	6 M HCl, 110°C, 1 h *or*	7–9
	5 M KOH, 115°C, 30 min.	9

The above references also contain methods for analysis of phosphoamino acids by cation-exchange, (6, 7); anion-exchange (10); ion-pair reverse-phase (7) HPLC and by two-dimensional TLC (6, 8).

and may be prefered for microanalytical work (see Section 3.2). For minimizing the destruction of certain amino acids, other hydrolysis conditions are given in *Table 1*.

More accurate estimates of serine, threonine and tyrosine content can be determined either by assuming 10% loss of serine and 5% loss of threonine and tyrosine over 24 h, or by performing 24, 48 and 72 h hydrolyses and extrapolating the results to zero time. The longer analyses are more likely to yield accurate results for valine, isoleucine and cysteic acid which can be released slowly from stretches of hydrophobic residues. For small peptides, hydrolysis for 16–18 h is generally sufficient.

Asparagine and glutamine are converted to their corresponding acids by acid hydrolysis and cannot, therefore, be quantified. However, prior treatment of the protein or peptide with bis (1,1-trifluoroacetoxy)iodobenzene (final concentration of 36 mg/ml in 0.01 M trifluoroacetic acid, TFA) for 4 h at 60°C in the dark converts the carboxamide moieties to their corresponding amines (11). Following dialysis against water, extraction with an equal volume of *n*-butyl acetate and lyophilization, the protein can be hydrolysed in 6 M HCl as above. The treatment liberates the amine derivatives of asparagine and glutamine which elute on ion-exchange systems near lysine.

Cysteine and cystine in proteins generally require modification before they can be quantified. The technique of *Table 2*, which gives maximum information, is recommended. Half the protein sample is reacted with 4-vinylpyridine at pH 8.5 to modify free-SH groups. To determine the disulphide content, the remainder is reduced under denaturing conditions and alkylated with the same reagent (all as described in Chapter 1, Table 10). Excess reagent is removed by de-salting, dialysis or organic solvent precipitation (Chapter 1). Hydrolysis and analysis of the two samples should now give free sulphydryl and total cysteine compositions respectively.

Alternatively, oxidation with performic acid converts all cysteine and cystine to cysteic acid and methionine to methionine sulphone.

(i) Add 0.5 ml of 30% (v/v) H_2O_2 to 4.5 ml of 88% formic acid, 25 mg phenol.

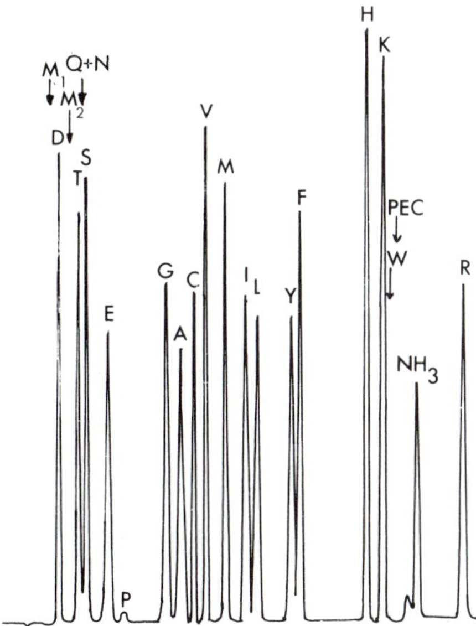

Figure 1. Elution of 10 nmol of amino acids (Pierce Chemical Co., standards 'H') from the resin column of an LKB alpha plus amino acid analyser. Elution positions of methionine sulphoxides (M_1), methionine sulphone (M_2), pyridylethylcysteine (PEC), tryptophan, glutamine and asparagine are also shown.

(ii) After 30 min at room temperature, cool to 0°C.

(iii) Add this reagent to the protein sample at 0°C to give a final protein concentration of 1%.

(iv) After 16–18 h, dilute the solution with an equal volume of water at 0°C and dialyse against two changes of 100 volumes of water at 4°C and finally against 100 volumes of 1 mM 2-mercaptoethanol at 4°C.

(v) The derivatized protein is lyophilized or dried in an hydrolysis tube ready for acid cleavage.

Homoserine and homoserine lactone arise from treatment with CNBr of protein and peptides containing methionine. TFA at 20°C for 1 h converts most of the homoserine to the lactone form which will elute after the NH_3 peak on the analyser.

Tryptophan destruction during acid hydrolysis can be minimized but not completely prevented by the methods given in *Table 1*. The tryptophan content can be verified by monitoring the decrease in absorbance at 280 nm following oxidation of its side chain with *N*-bromosuccinimide (NBS). The protein is dissolved in 50 mM acetate buffer pH 4.0 at a concentration of up to 2×10^{-4} M. Aliquots of NBS solution in the same buffer are added until there is no further decrease in absorbance after about 20 min which cannot be accounted for by dilution alone. The extinction due to tryptophan in the sample solution is estimated by multiplying the decrease in absorbance at 280 nm by the empirical factor 1.31 and using the extinction coefficient for tryptophan at 280 nm of 5500.

1.1.2 *Chromatography*

Figures 1 and *2* show the elution order (see also *Table 3*) of amino acids and derivatives

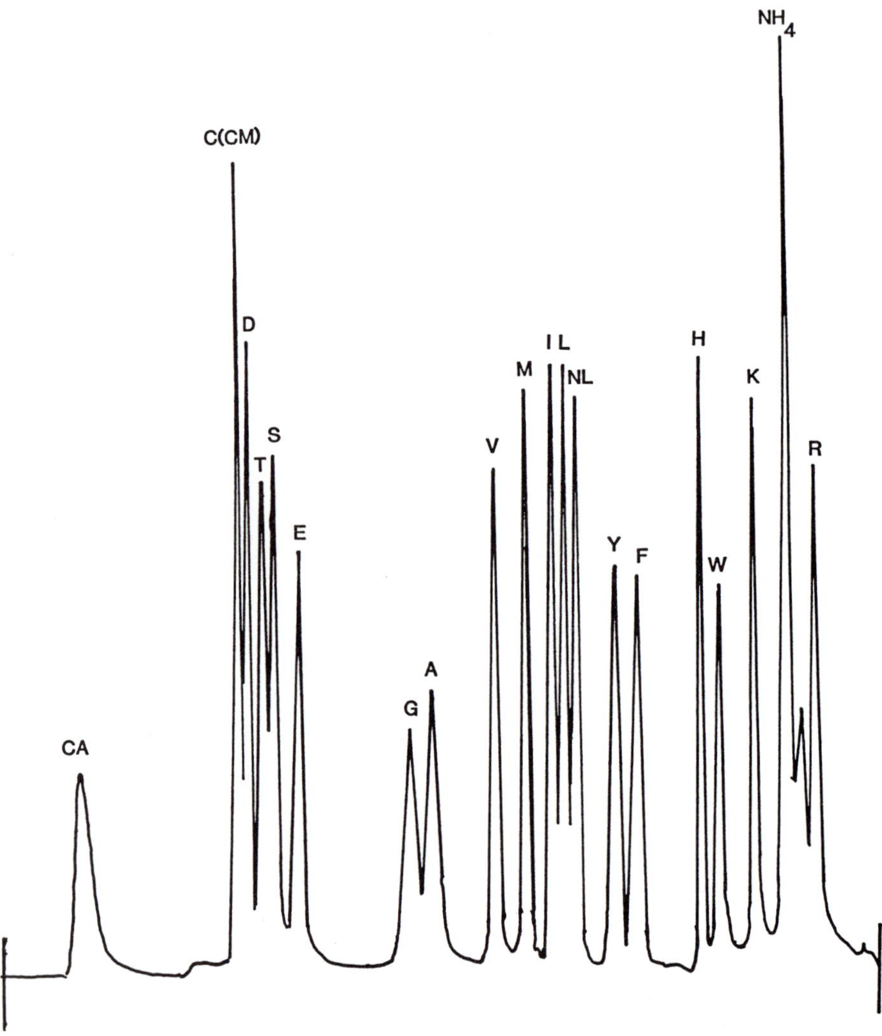

Figure 2. Elution of amino acids (BDH) on Rank-Hilger Chromaspek analyser using a sulphonated polystyrene resin and a continuous gradient profile. Starting buffer: citric acid (10.5 g), NaCl (11.7 g), thiodiglycol (2.5 ml) to 1 litre, pH 2.25. Terminating buffer: Tris–Na citrate (14.7 g), Na_3PO_4 (6.46 g), 4% EDTA (12.5 ml) to 1 litre, pH 11.5.

Table 2. Synthesis of S-β-(4-pyridylethyl)-L-cysteine (17).

1.	Dissolve 500 mg of L-cysteine in de-ionized water and add 0.57 ml (4 mmol) of triethylamine followed by 0.43 ml (4 mmol) of 4-vinylpyridine (Aldrich; redistil before use).
2.	Stir the mixture under N_2 for 24 h and dry at 40°C in a rotary evaporator.
3.	Re-crystallize three times from a concentrated aqueous solution using ethanol as a precipitant and leaving to stand in the cold.
4.	The needle-shaped crystals can be recovered by filtration and dried *in vacuo* to constant weight. The compound is stored at room temperature.

Table 3. Elution order of amino acids and derivatives from ion-exchange analysers.

1. *o*-Phosphoserine	23. S-Ethylcysteine
2. *o*-Phosphothreonine	24. Glucosamine
3. Cysteic acid	25. Mannosamine
4. Urea	26. Galactosamine
5. Glucosaminic acid	27. Valine
6. Methionine sulphoxides	28. Cysteine
7. Hydroxyproline	29. Methionine
8. Aspartic acid	30. α-methylmethionine
9. Methionine sulphone	31. Isoleucine
10. α-methyl aspartic acid	32. Leucine
11. Threonine	33. Norleucine
12. Serine	34. Tyrosine
13. Asparagine	35. Phenylalanine
14. Glutamine	36. Ammonia
15. α-methyl serine	37. Hydroxylysine
16. Homoserine	38. Lysine
17. Glutamic acid	39. 1-Methylhistidine
18. α-methyl glutamic acid	40. Histidine
19. Proline	41. 3-Methylhistidine
20. S-methylcysteine	42. Tryptophan
21. Glycine	43. Homocysteine thiolactone
22. Alanine	44. Arginine

from the column of an LKB alpha plus amino acid analyser and the Rank-Hilger Chromaspek respectively. The elution conditions are given in the respective legends; the first relies on a step gradient and the second on a continuously changing gradient. The detection system generally employs ninhydrin but, for increased sensitivity OPA can be used and the resulting fluorescence detected.

1.2 High-performance liquid chromatography

HPLC techniques have the potential for rapid, routine and reproducible amino acid analysis. Reverse-phase systems show much promise and a few are described below. Generally, pre-column derivatization of the amino acids is necessary because the background in the buffers limits post-column methods. One of the most popular reagents, OPA (12,13), suffers the twin disadvantages of derivative, particularly lysine, instability and lack of reaction with proline or hydroxyproline unless preoxidation with Chloramine T is carried out. Double derivatization methods have been used, where a second pre-column procedure following OPA reaction (e.g. FMOC) is carried out to react proline (14). The advantage of the OPA method is its sensitivity and both Waters and Hewlett Packard have published methods based on this reagent.

In this section, however, the derivatization of amino acids with phenylisothiocyanate (PITC) to yield the phenylthiocarbamyl (PTC)-derivative is described in detail because it is felt that this method promises widest utility in the future and requires a minimum of special equipment. The reader is referred to reference 15 for other details of modern high-sensitivity HPLC amino acid analysis such as cation exchange of free amino acids and reverse-phase determination of OPA-, dansyl-, dabsyl- (a promising method has been standardized by Ciba-Corning, Cambridge, UK), and FMOC-derivatives. Two automated systems for amino acid composition determination using PITC are available commercially. These are the PICOTAG system of Waters/Millipore Corp. and the model 420 derivatizer/130A analyser system of Applied Biosystems Inc.

1.2.1 *Sample hydrolysis*

Protein or peptide samples must be substantially free of salt, buffers and detergents for good results. The most universally acceptable way to achieve this is by de-salting into volatile solvent/buffer systems (see Chapter 1, Section 4). Buffer salts in particular may give rise to problems by keeping the pH too low for complete reaction of PITC with primary and secondary amines in the sample. The purity of all reagents and cleanliness of surfaces in contact with the samples is essential. Heavy metal contamination leads to low recoveries of certain PTC-amino acids, especially that of lysine. Accurate results on quantities below 50 pmol of each amino acid will prove difficult to obtain, unless vapour phase hydrolysis (described below) is employed, together with stringent cleanliness and minimal time lapse between derivatization and analysis.

The de-salted sample ($0.1-10$ μg) is collected in a small hard glass tube pre-cleaned by heating in 6 M HCl and/or pyrolysis at 500°C (annealing furnace). For identification, tubes should be numbered near the bottom with a diamond scriber. The volatile sample buffer should be removed *in vacuo*, either in a dedicated desiccator or a vacuum centrifuge to reduce the risk of loss or contamination.

Hydrolysis is performed most conveniently (especially for large numbers of samples) in screw-top glass or PTFE (Tuftainer; Pierce) vessels containing the hydrolysis acid.

(i) Place the numbered tubes in the hydrolysis vessel containing $0.5-1.0$ ml of 6 M HCl and 0.5% phenol (v/v).
(ii) Flush the vessel with 99.998% argon (sequencer grade) for 2 min and close with a Teflon disc bonded to silicon rubber and a heat-resistant screw cap (all available from Waters, ABI or Ciba-Corning).
(iii) Heat the reaction vessel in an oven or heating block at 110°C for 24, 48 or 72 h (16). Alternatively, hydrolysis at 165°C for 1 h may be used (*Table 1*).

Addition of $1-2$ mM phenol assists the recovery of tyrosine. For acid-labile amino acids, especially tryptophan, hydrolysis conditions are shown in *Table 1*. To minimize condensation of HCl in the tubes after removal from the oven, the vials can be opened *very slowly* while still hot, provided the operator wears protective gloves and glasses and works in a fume hood. After cooling, the vials are placed *in vacuo* in a desiccator containing sodium hydroxide pellets to remove excess acid. If analysis is delayed, the samples should be stored dry at room temperature.

Determination of cysteine is problematic in PTC-derivatization, since several commonly used derivatives are difficult to resolve in most separation systems. PTC-cysteic acid migrates with the solvent front on C18 columns and carboxymethyl cysteine migrates very close to other polar PTC-amino acids. An acid derivative which is well resolved in most separation systems is 4-pyridylethyl cysteine. Further advantages of this modification are that the PTH-derivative is well resolved in sequence analysis and cysteine-containing peptides can be detected by the absorbance of the pyridylethyl group at 256 nm. A routine method for pyridylethylation of proteins is given in Table 9, Chapter 1 (17).

1.2.2 *Derivatization with PITC*

The following stages of the procedure should also be carried out on the amino acid standards, such as Pierce mixture H (2.5 μmol/ml in 0.1 M HCl). 25 nmol (10 μl) of these standards should be dried, along with the hydrolysed samples obtained from the preceding step. A reference amount of 4-pyridylethyl cysteine can be added if necessary (synthesis in *Table 2*).

If the derivatization is being carried out manually, residual acid from the hydrolysis must be neutralized or completely removed by drying *in vacuo*. Detailed results and assessment of derivatization of amino acids with PITC can be found in references (18–20).

(i) Add 100 μl of coupling buffer (0.7 ml of acetonitrile, 0.2 ml of triethylamine, 0.1 ml of water—stored at −20°C for no longer than 1 month) to each tube

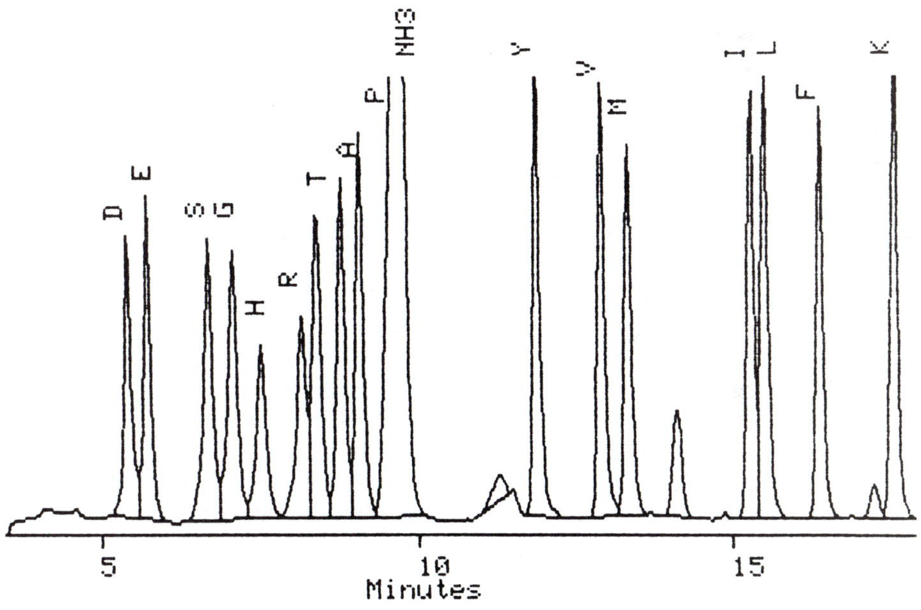

Figure 3. Elution of 50 pmol PTC-amino acids from a Brownlee octadodecylsilane PTC column. Column: ABI PTC C18 220 × 2.1 mm; wavelength: 254 nm; temperature: 38°C; flow: 0.3 ml/min; sample buffer: 50 mM sodium acetate pH 5.0; solvent A: 50 mM sodium acetate pH 5.4; solvent B: 70% (v/v) CH_3CN/water.

Gradient conditions:

Time (min)	*% B*
0	7
10	32
20	55
25	100
30	100
31	7

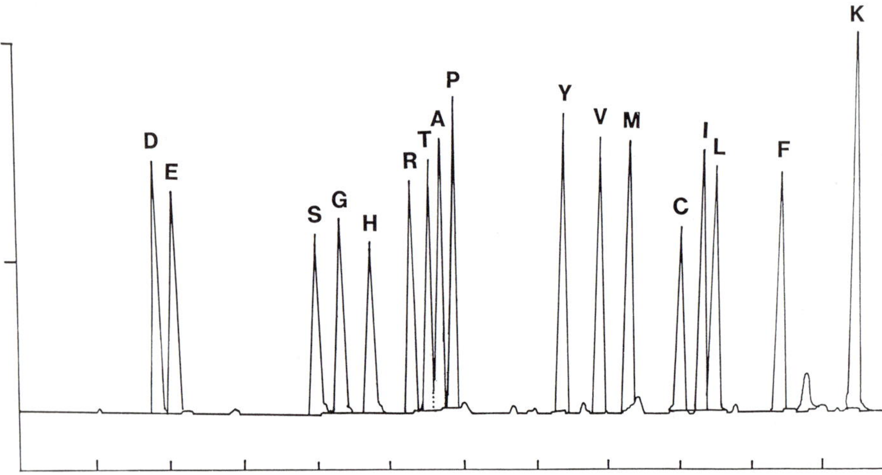

Figure 4. Elution of 100 pmol of amino acids using a PICOTAG system. Column: Pico-Tag (or NovaPak C18, 60-100 Å, 4 μm, 3.9 × 150 mm); flow-rate: 1.0 ml/min; wavelength: 254 nm; temperature: 40°C; injection volume: 20 μl (out of 100 μl); solvent A: 139 mM CH_3COOH, 3.6 mM triethylamine (TEA), pH 6.5/CH_3CN (94/6); solvent B: 60% CH_3CN.

Gradient conditions:

Time (min)	*Flow*	*% A*	*% B*	*Curve*
Initial	1.0	100	0	*
10.5	1.0	54	46	5
11.0	2.0	0	100	6
13.0	2.0	0	100	6
13.5	2.0	100	0	11
20.0	1.0	100	0	11

containing the dried amino acid samples, vortex and remove by vacuum centrifugation.

(ii) Add a further 100 μl of coupling buffer together with 5 μl of PITC, vortex and incubate at room temperature for 20 min. Dry in a vacuum centrifuge until a buff of yellowish deposit (mainly thioureas) remains in the tubes. If an analysis cannot be performed immediately, the samples are best stored dry at −20°C under argon.

1.2.3 *HPLC analysis of PTC-amino acids*

Application-specific columns for PTC analysis are available from Waters/Millipore Corp., and Applied Biosystems Inc. (ABI) together with separation protocols and solvent systems (*Figures 3* and *4*). Many workers have also had success with Altex ultrasphere octadecylsilane and Zorbax Bio Series PTH, reverse-phase columns.

2. LOCATION AND ISOLATION OF PARTICULAR PEPTIDES OR PEPTIDES CONTAINING SPECIFIC AMINO ACIDS

Identification of peptides which contain particular amino acids can be very helpful in sequence analysis for cDNA cloning, and also in biotechnology for assessing processing of the expressed protein. The most important methods are UV absorption, radiolabelling and chemical modification (effectively an artificially-induced change in chromatographic behaviour). HPLC is generally the most useful purification method to be applied in association with these techniques and is fully described in Chapter 1. The older methods of using specific amino acid stains by dipping or spraying thin layer chromatograms on which peptides have been separated by electrophoresis or ascending solvent chromatography may still be useful and are described here for completeness.

2.1 **Detection of specific amino acids in peptides**

2.1.1 *Amino groups*

A convenient, rapid and quantitative test is achieved with 2,4,6-trinitrobenzenesulphonate (TNBS).

(i) Dissolve the protein or peptide in 50 mM Na_2CO_3 pH 9.5, place 1 ml in a spectrophotometer cuvette and add 100 μl of TNBS (7.2 mg/ml in water).
(ii) Monitor at 367 nm (reaction may take some time to come to completion).
(iii) The extent of trinitrophenylation can be calculated using an extinction coefficient at 367 nm of 1.1×10^4 M^{-1} cm^{-1} (21).
(iv) For detection of peptides on thin layers, dip the plate into a 0.2% (w/v) solution of ninhydrin in ethanol: acetic acid (20:1 v/v) and dry at 60°C for 30 min (oven).
(v) For increased sensitivity, dry the thin layer plate, then dip into 2% (v/v) triethylamine in dry acetone, air dry for several minutes, then dip into fluorescamine (0.01% w/v) in dry acetone.
(vi) After drying, stained peptides are detected with long-wave UV light (336 nm).

2.1.2 *Glycopeptides (O- or N-linked)*

Carbohydrate linked to polypeptides can be detected in solution with about 5–10 nmol of polypeptide.

(i) Collect the peptide into borosilicate glass tube and dry.
(ii) Set up a series of glucose standards (1–100 μg) in a similar way.
(iii) Add 0.1 ml of 5% (w/v) aqueous phenol to the tubes and vortex.
(iv) During vortexing, slowly add 0.5 ml of concentrated H_2SO_4.
(v) After 20 min measure the absorbances at 490 nm.

2.1.3 *Arginine (phenanthrenequinone; ref. 22)*

(i) Dip the paper or spray the TLC with a solution of equal volumes of 0.02% (w/v) phenanthrenequinone in anhydrous ethanol (store at 4°C in the dark) and 10% (w/v) NaOH in 60% (v/v) ethanol.
(ii) Air-dry for 20–30 min, and visualize with a UV lamp (366 nm) when arginine-containing peptides appear as fluorescent spots.

2.1.4 *Histidine*

The classical method involves the use of the Pauly reagent [1 g of sulphanilic acid in 10 ml of 1 M HCl mixed with an equal volume of 0.7% (w/v) aqueous $NaNO_3$]. Dip the plate (or spray), dry in an air stream and spray lightly with 10% (w/v) aqueous Na_2CO_3. Histidine peptides show up as red/orange spots.

2.1.5 *Tryptophan (Ehrlich)*

Immediately before use mix one part of concentrated HCl with nine parts of 2% (w/v) *p*-dimethylaminobenzaldehyde in acetone. Surface dip the TLC plate. Tryptophan-containing peptides give purple colouration. The test may follow ninhydrin detection of peptides. Tryptophan-containing peptides are also indicated by high absorption at 279 nm.

2.1.6 *Tyrosine*

Surface dip or spray the TLC plate in a 0.1% solution of α-nitroso-β-naphthol in 75% alcohol, air-dry and treat again with 10% aqueous HNO_2. Heat at 90°C for 3 min when tyrosine-containing peptides will appear red on a pale green background.

2.2 Determination of disulphide bonds

Classical techniques for determining disulphide bond linkage patterns usually require the fragmentation of proteins into peptides under low pH conditions to prevent disulphide exchange. Pepsin or cyanogen bromide are useful for this [Sections 4.1.4(ii), 4.2.6]. After electrophoresis at pH 6.5 the TLC plate is exposed to performic acid which oxidizes cystine residues to cysteic acid. Electrophoresis in the second dimension then produces off-diagonal spots which represent two peptides previously covalently-linked. These techniques are discussed by Brown and Hartley (23,24). Modern micromethods rely heavily upon HPLC and a suggested strategy is given below.

(i) Alkylate the protein to prevent possible disulphide exchange by dissolving in 100 mM Tris–HCl pH 8.5 and adding 1 μl of 4-vinylpyridine. Incubate for 1 h at room temperature and de-salt by HPLC or precipitate/extract with 95% ice-cold ethanol.

(ii) Fragment the protein under conditions of low pH (e.g. pepsin, CNBr) and subject the peptides from half the digest to HPLC. Both conditions or other enzymes may be required if fragments are large.

(iii) To the other half of the digest (dried and re-suspended in 10 μl of isopropanol) add 5 μl of 1 M triethylamine–acetic acid pH 10, 5 μl of 1% tri-*n*-butyl-phosphine in isopropanol and 5 μl of 4-vinylpyridine. Incubate for 30 min at 37°C, and dry *in vacuo*, re-suspending in 30 μl of iso-propanol twice. This procedure cleaves the disulphides and modifies the resultant -SH groups.

(It is also possible to react the reduced material with radioactive iodoacetamide to ease detection of the disulphide-linked cysteines—see Appendix).

(iv) Cysteine-linked peptides are identified by the differences between reduced and unreduced samples as revealed by HPLC. Collection of the alkylated peptides (which can be identified by re-chromatography at 254 nm) and sequence analysis should allow disulphide assignments to be made.

2.3 Selective isolation by hydrophobicity modulation

A number of methods permit the identification of peptides by shifts in their relative hydrophobicity detected by elution of normal and modified forms from reverse-phase resins. The modifications may be produced *in vivo* or may have been introduced *in vitro* prior to the analysis. This type of technique is very useful for detection of post-translational modifications or point mutations in proteins of known primary structure, and for rapidly locating regions of a protein produced by expression of recombinant DNA.

2.3.1 *Prediction of peptide retention times on reverse-phase HPLC*

The original ideas were developed by Meek (25) and further elaborated by others (26,27). Essentially the method involves calibration of reverse-phase octyl or octadecyl silyl HPLC columns with a standard linear gradient and peptides of known sequence and/or composition. For peptides of up to about 40 residues, there is a high correlation between elution time (or solvent composition) and the sum of the retention coefficients derived from the component amino acids for a given reverse-phase system. Peptides of known sequence/composition which elute significantly outside the expected 'windows' calculated from the sum of the retention coefficients are likely to be modified in some way.

A suitable group of peptides for calibrating a reverse-phase column can be obtained from a tryptic digest of a known protein (myoglobin or β-lactoglobulin). The observed retention times are plotted against ln $(1 + i\Sigma n_i{*}c_i)$ where c_i is the coefficient of amino acid i and n_i is the number of times which it occurs. *Table 4* gives a starting set of values for use with a C_{18} μBondapak column operated with a gradient of 0–40% acetonitrile in 0.1% TFA over 2 h at a flow-rate of 1.5 ml/h. The reader is referred

Table 4. Retention coefficients on C18 reverse-phase HPLC.

Amino acid or substituent	*Retention coefficient*
Leu	20.0
Phe	19.2
Trp	16.3
Ala	7.3
Ile	6.6
Tyr	5.9
Met	5.6
Pro	5.1
Val	3.5
Thr	0.8
Gln	−0.3
Gly	−1.2
His	−2.1
Asp	−2.9
Arg	−3.6
Lys	−3.7
Ser	−4.1
Asn	−5.7
Glu	−7.1
Cys	−9.2
$-CO_2H$	2.4
NH_2-	4.2
N-acetyl−	10.2
Amide	10.3

to Meek (25) for details of adjusting the individual retention coefficients. Anomalies which may cause column to column variations or aberrant elution by some unmodified peptides may be reduced by addition of 0.1% triethylamine to the eluting buffers and readjusting the pH appropriately with TFA. Residual ion-exchange effects between histidine, lysine and arginine residues and the column matrix may also be suppressed in this way.

2.3.2 *Histidine-containing peptides*

The hydrophobicity of these peptides can be altered by ethoxyformylation (28).

(i) Re-dissolve dried enzyme digests (trypsin, chymotrypsin, etc.) in 120 μl of 20% (v/v) isopropanol−water.

(ii) Add 1 μl of ethoxyformic anhydride (Sigma) in ethanol used at a concentration so as to give at least a 20-fold molar excess of reagent over histidine.

(iii) After 30 min at 25°C, subject the mixture to, for example, reverse-phase HPLC (or freeze at −20°C).

(iv) Dry under vacuum peptides absorbing at 242 nm (His- and Tyr-containing); re-dissolve in 150 μl of 0.3% TFA and incubate at 55°C for 1 h.

(v) Subject the peptides to re-chromatography. Histidine-containing peptides, now desethoxyformylated, will elute earlier in the same gradient. The elution shift also constitutes an effective purification step.

Similar strategies can be used for the modification of methionine (29) and tryptophan-containing peptides (30).

2.3.3 *Phosphoserine-containing peptides*

Reversible phosphorylation at serine, threonine or tyrosine residues is a principal mechanism for regulating the function of proteins, particularly in response to extracellular signals, and most cellular processes in eukaryotes are controlled by this post-translational modification (31). Rapid and sensitive methods for identifying and purifying phosphopeptides could therefore, be very useful.

Meyer *et al.* (32) showed that the phosphoseryl derivative of the peptide LRRASLG could be converted to its S-ethylcysteine derivative by β-elimination of the phosphoseryl residue in NaOH and addition of ethanethiol. The procedure is not applicable however, to phosphothreonine- and phosphotyrosine-containing peptides.

The conversion of phosphoserine to S-ethylcysteine (which increases hydrophobicity) has been exploited to provide a simple procedure for the selective isolation of phosphoserine-containing peptides from complex mixtures (33). Following initial HPLC separation of a proteolytically digested protein, containing either ^{32}P-labelled or 'cold' phosphoserine, phosphorylated peptides can be identified by the presence of radioactivity or fast atom bombardment mass spectrometry (34,35). Subsequent modification with ethanethiol selectively increases the hydrophobicity of phosphoserine-containing peptides thereby changing their chromatographic behaviour. The number of S-ethylcysteine residues in a peptide can be quantitated using their PTC-derivative. The procedures described are particularly powerful for the analysis of peptides that are phosphorylated at multiple sites, since it is possible to completely resolve mono-, di- and even tri-phosphorylated forms of the same peptide by reverse-phase HPLC (35).

It is advisable to standardize the procedure using a commercially available synthetic peptide such as LRRASLG (Sigma) which is a good *in vitro* substrate for cAMP-dependent protein kinase.

All manipulations should be carried out in a fume hood, as ethanethiol (Pierce-Warriner) is volatile (boiling point, 35°C) and has a very pungent odour. All reagents should be flushed for 10 min with N_2 (ethanethiol is best kept on ice at this time) prior to use.

(i) Dissolve phosphoserine-containing peptides (0.2–2 nmol) in 50 μl of a reaction mixture consisting of water (200 μl) dimethylsulphoxide (200 μl) ethanol (100 μl), 5 M NaOH (65 μl) and ethanethiol (60 μl). Incubate in screw-top vials for 1 h at 50°C under N_2.

(ii) Terminate incubation by cooling in a dry ice/isopropanol bath and add 10 μl of glacial acetic acid (Aristar grade). After dilution to 1 ml with water and re-freezing, dry the reaction products, preferably on a vacuum concentrator.

(iii) Dissolve the reaction products from the previous step in 0.1% (v/v) aqueous TFA and re-fractionate by reverse-phase HPLC under the same conditions as those prior to chemical modification. Peptides containing ethylcysteine will now be selectively retarded on the reverse-phase column and elute at significantly higher acetonitrile concentrations. This procedure is effectively demonstrated by chromatography of the phosphorylated and dephosphorylated forms of the peptide SPQSRRSESSEE following incubation with ethanethiol in the presence of NaOH (33).

(iv) Since S-ethylcysteine is stable to acid hydrolysis in 6 M HCl *in vacuo* for 24 h at 110°C, the number of phosphoserine residues in a peptide can be quantified by amino acid analysis. This is conveniently performed by pre-column derivatization of hydrolysed amino acids to their PTC-derivatives with 20 μl PITC:ethanol:water:triethylamine (1:7:1:1 by vol.). Dry the PTC-derivatives in a vacuum concentrator and analyse by reverse-phase HPLC, as described earlier. The PTC-derivative of S-ethylcysteine elutes between PTC-methionine and PTC-isoleucine using the Waters Associates 'PICO-TAG' C18 reverse-phase system. Typical recoveries for S-ethylcysteine are similar to those for methionine.

(v) The location of the S-ethylcysteine can be determined by automated sequencing.

(vi) This procedure can also be applied to a peptide containing more than one site of phosphorylation, for example, EQESSGEEDSDLSPEER in protein phosphatase inhibitor-2 which contains two phosphoserine residues as judged by fast atom bombardment mass spectroscopy (35). Complete purification of the peptide was achieved using modification with ethanethiol and subsequent rechromatography on a Vydac C18 column. Amino acid analysis as above of the modified peptide confirmed that two of the four serine residues had been converted to S-ethylcysteine. Gas-phase sequencing established residues 4 and 5 as the phosphorylated serines.

3. RADIOLABELLING PROTEINS FOR PEPTIDE MAPPING

Radioisotopic labelling of proteins is of great value in their purification and characterization. Often, they can be labelled *in vivo* in tissue culture or in cell-free homogenates, prior to isolation. In such cases, they are most easily recovered following sodium dodecyl sulphate–polyacrylamine gel electrophoresis (SDS–PAGE) autoradiography and electroelution (see Chapter 1, Table 1, and Chapter 3).

However, unlabelled protein can be iodinated *in situ* in polyacrylamide gel slices. Many applications require a peptide map and this can be obtained by digestion of the labelled peptide *in situ* in the slice with trypsin, followed by two-dimensional mapping and autoradiography.

3.1 **Iodination of protein in gel slices**

(i) Place the gel slices (stained or de-stained) in 'Eppendorf' tubes multiply perforated with a hot needle. Seal and incubate these tubes in 1 litre of 25% isopropanol for 8 h followed by 1 litre of 10% methanol for 16 h.

(ii) Transfer to fresh Eppendorf tubes, dry under vacuum and add 10–20 μl of [^{125}I]KI (500 μCi). After the liquid has been taken up by the gel add 50 μl of 0.5 M sodium phosphate pH 7.5.

(iii) Add 10 μl of chloramine T (5 mg/ml) and incubate for 30 min at room temperature. Then add 1 ml of metabisulphite (1 mg/ml) and incubate for 15 min at room temperature.

(iv) Wash 10 times with 1 ml of 10% methanol followed by incubation in perforated 'Eppendorfs' against 2 × 1 litre of 10% methanol. 0.2 g of mixed bed ion-exchange resin in dialysis tubing will remove some of the ^{125}I in the dialysate.

(v) When no more ^{125}I is being dialysed out, transfer the gel slices to 'Eppendorf' tubes and dry as before.

3.2 **Digestion with trypsin**

(i) Add 0.5 ml of 0.05 M NH_4HCO_3 to the gel slice and incubate at 37°C for 24 h with 30 μg of trypsin.

(ii) Remove the supernatant containing the eluted peptides, freeze the samples at −70°C and lyophilize.

3.3 **Two-dimensional mapping**

(i) Apply 1–5 μl of peptide in acetic acid: formic acid: water (15:5:80 by vol) to cellulose TLC plates (E.Merck, cat. no. 5577-7 or CEL 400, Macherey–Nagel), 2 cm from each edge at the lower left hand corner.

(ii) Turn the plate through 180°C and apply 2 μl of tracking dye (2% orange G + 1% acid fuchsin in above solvent), 2 cm from left hand edge and 1 cm from bottom of plate.

(iii) Mark a line 2 cm long, 10.5 cm from the bottom left hand edge of the plate.

(iv) Place the plate on the cooling surface of a flatbed electrophoresis apparatus with the dye spot towards the negative electrode and the indicator mark between the two electrodes. Fill the buffer chambers, moisten a filter paper disc in buffer and use it to dampen the areas around the sample and dye spots so that buffer flows to concentrate these. With a tissue, dampen the remainder of the plate and overlay with a clean 20 cm × 20 cm glass plate.

(v) Electrophorese at a constant 1000 V (5–12 mA) until the large tracker dye spot migrates to the indicator mark at 10.5 cm. Remove the plate to a well-equilibrated glass chromatography tank with a good seal.

(vi) Ascending chromatography in *n*-butanol:pyridine:acetic acid:water (6.5:5:1:4 by vol.) takes about 8 h. Remove the plate and air-dry for 2 h in a fume hood. Place the plate in a light-tight cassette after covering it with cling flim or grease-proof paper. Expose with Kodak X-Omat XRP-5 (blue) film. Spot sharpness is enhanced by over-exposing the film and under-developing (stop development rapidly with a bath of 1% acetic acid prior to fixing).

4. CLEAVAGE PROCEDURES FOR PROTEINS AND PEPTIDES

4.1 Enzymatic methods

The conditions used for the cleavage of proteins by proteases can be varied depending on the results required. Traditionally complete digestion was often sought but with the advent of automatic sequencing equipment there has been a tendency to move towards more limited incubation conditions which would generate fewer and larger protein fragments. It is worth therefore making a few general points before describing particular reaction systems.

(i) Cleavage of the polypeptide chain is greatly facilitated if the protein is denatured; native proteins present far fewer immediately accessible cleavage sites.

(ii) Buffers such as NH_4HCO_3 or *N*-ethylmorpholine are preferred since they can mostly be removed by lyophilization.

(iii) The time, temperature and enzyme/substrate ratio of the reaction can all be varied to generate a desired fragmentation pattern.

(iv) The reaction may be terminated by acidification, freeze-drying, rapid boiling or the addition of inhibitors such as phenylmethylsulphonyl fluoride (PMSF, serine proteases) iodoacetate (thiol proteinases) or EDTA (metalloproteinases).

(v) It is often useful to follow the progress of the digestion using SDS−PAGE.

4.1.1 *Acidic residues*

(i) *Staphylococcus aureus (V8) protease (EC 3.4.21.19).* This useful protease (also called endoproteinase Glu-C) is essentially specific for peptide bonds on the C-terminal side of glutamate residues, with some cleavage occurring at aspartyl bonds (36,37). A few non-specific cleavages can take place under harsh conditions. Cleavage is inhibited by bulky side chains in the $n+1$ position. The enzyme is active in the pH range 3.5−9.5 (maximum 7−8) and its specificity may be restricted almost completely to glutamyl residues by the pH and buffer conditions employed.

(1) For cleavage largely at glutamate residues, dissolve the protein in 0.1 M ammonium bicarbonate pH 8.1 at a concentration of about 10 mg/ml and add 2% (w/w) enzyme to substrate. Incubate at 30−37°C for 2−3 h. Terminate the digestion as described for trypsin.

(2) To extend susceptibility to aspartyl bonds, perform the incubation with enzyme in 0.1 M sodium phosphate buffer pH 7.8. In this case the digests should be de-salted prior to HPLC. However, cleavage at aspartate residues is much less efficient than at glutamate residues, even with denatured proteins.

In some cases good cleavage of proteins may require extended digestion times (12−48 h) particularly if the sample is not completely soluble in the reaction buffer. 4 M urea or 0.2% SDS may be used as an aid to solubility, since the enzyme is active under these conditions.

This protease can be very successfully employed with native proteins (both soluble and membrane-bound) as a means of generating larger fragments resulting from cleavage at a few susceptible sites. The enzyme cleaved rhodopsin for example at only two bonds when incubated in 0.67 M phosphate for 2−3 h at room temperature (38).

(ii) *Endoproteinase Asp-N.* This enzyme is a metalloproteinase (Boehringer) which cleaves peptides bonds N-terminal to aspartic and cysteic acid residues (39).

(a) Dissolve protein/peptides in 50 mM sodium phosphate pH 8.0 (up to 1 M urea, 0.001% SDS and 1 M guanidine−HCl can be added to ensure solubility).
(b) Incubate with up to 5% (w/w) enzyme dissolved in the phosphate buffer for 2−18 h at 25−37°C.

4.1.2 *Basic residues*

(i) *Trypsin (EC 3.4.21.4).* Trypsin is still the most widely used proteolytic enzyme and cleaves proteins specifically on the carboxy side of arginine, lysine and S-aminoethylcysteine residues. Anomalous cleavage may occur on prolonged digestion, particularly on the carboxyl side of hydrophobic (especially tyrosine) residues. These are often attributed to contaminating chymotryptic activity even in ψ-trypsin treated with TPCK, but may in fact be due sometimes to an inherent trypsin activity. Poor, or no, cleavage at Arg-Pro and Lys-Pro bonds are usual.

(a) Dissolve the protein in 0.2−0.5 M *N*-ethylmorpholine−HCl (NEM buffer) pH 8.5 or in 0.1 M NH_4HCO_3 pH 8.1 at a concentration of up to 20 mg protein/ml. Add enzyme solution (freshly prepared in the same buffer or in 1 mM HCl when it may be stored frozen at −20°C) to give a ratio of 2% (w/w) trypsin/protein and incubate with constant or intermittant agitation for up to 4 h at 37°C. Alternatively the trypsin can be added in 1% portions separated by hourly intervals. Extended incubations (6−7 h) are not profitable since the enzyme 'autodestructs'.
(b) Terminate small digests (1 mg) by freezing (liquid air, liquid N_2, alcohol bath) and lyophilizing. Re-dissolve in 100 μl of 0.1% TFA and purify peptides directly by reverse-phase HPLC as described elsewhere.
(c) Apply large digests directly onto a 1 × 200 cm column of Bio-Gel P6 in 0.05 M NH_4HCO_3 pH 8.1 (after removing any insoluble material by centrifugation). Monitor the eluate at 220 and 280 nm, collect major peaks and lyophilize. Purify peptides from each peak by HPLC.

The specificity of trypsin can be restricted to arginine residues by modifying the lysines by citraconylation, maleylation or succinylation (see Section 6.2, Chapter 6) or can be restricted by lysine residues, though perhaps rather less successfully, by reacting arginines with 1,2-cyclohexanedione (40) or *p*-hydroxyphenylglyoxal (41, see also Appendix). If such specificity is desired, less problems are encountered if arginine- or lysine-specific enzymes are employed (see below). Since trypsin is also active in up to 4 M urea, proteins which are not soluble in the above buffers may be dissolved in a small volume of 8 M urea and diluted before addition of the protease. Alternatively, a fine suspension of an insoluble protein (produced for example by sonication) can be digested by the same enzyme. In this case, higher levels of trypsin and more extended incubation times should be generally employed.

(ii) *Arginine-specific proteases*. A number of enzymes specific for Arg-X bonds have been used as an alternative to trypsin with substrates in which the lysine residues have been blocked. Of these, clostripain (42), thrombin and mouse sub-maxillary gland enzyme (endoproteinase Arg-C) (43) are commercially available (Boehringer).

(a) Clostripain (EC 3.4.22.8) is a sulphydryl protease and appears to require activating by pre-incubation in 1 mM calcium acetate, 2.5 mM dithiothreitol (DTT) for 2–4 h at 25°C (44). Incubate the protein with 1–2% (w/w) clostripain/protein in 25 mM phosphate buffer, pH 7.5, 0.2 mM calcium acetate, 2.5 mM DTT. A reaction time of 2–3 h at 37°C is generally sufficient to cleave most Arg-X bonds. Some authors have reported using overnight digestion but this may markedly increase non-specific cleavage. The enzyme is active in the presence of up to 6 M urea.

(b) The proteases from mouse sub-maxillary gland and thrombin are used in bicarbonate buffer, as described for trypsin, but with the incubation period extended to 6–8 h. Thrombin, in particular, may yield a very restricted cleavage pattern and is useful for generating a few large fragments.

(iii) *Lysine-specific enzymes*. Proteases specific for both X-lys and Lys-X bonds have been applied in sequence analysis.

(a) The enzyme from *Armillaria mellea* (EC 3.4.99.32) (45) cleaves X-Lys bonds with some minor cleavage at Arg-X. Dissolve the protein substrate in 0.2 M *N*-ethylmorpholine-acetate buffer, pH 7.5. Add 1% (w/w) enzyme and incubate under N_2 at 37°C for 24 h. Terminate the reaction as described for trypsin. If lysine residues are blocked by trifluoroacetylation and cysteine residues are converted to 2-aminoethyl derivatives, then this enzyme is reported to cleave specifically on the C-terminal side of the modified cysteine (46).

(b) Lys-X bonds may be cleaved by an enzyme from *Lysobacter enzymogenes* (endoproteinase Lys-C) in 0.1 M NH_4HCO_3 as described for trypsin (47). Some minor non-specific cleavage has been reported, notably at Asn-X bonds. The enzyme is still active up to 0.5% SDS. It can be used successfully with native protein to produce cleavage at only a few specific sites.

4.1.3 *Cleavage at hydrophobic residues*

(i) *Chymotrypsin (EC 3.4.21.1)*. This enzyme exhibits much broader specificity than trypsin, cleaving on the C-terminal side of hydrophobic residues, especially phenylalanine, tryptophan, tyrosine and leucine in order of decreasing susceptibility. Significant cleavage is sometimes found on the carboxyl side of methionine and histidine, particularly with extended incubation periods. Neighbouring residues can influence the rates of cleavage and peptide bonds involving proline appear to be relatively resistant. The method given above for digestion with trypsin also works well for chymotrypsin. Although incubation for 2 h normally gives complete cleavage of the most susceptible bonds, periods of 12–24 h may be used if the maximum number of small peptides is required.

(ii) *Thermolysin (EC 3.4.24.4) (48)*. This enzyme has broad specificity for the N-terminal side of hydrophobic residues, particularly leucine, isoleucine, phenylalanine and valine in descending order of preference. The enzyme is unresponsive to tryptophan, and proline residues appear to inhibit cleavage when attached to the C-terminal side of the hydrophobic residue. Thermolysin cleaves polypeptides more efficiently than peptides. The reaction conditions given above for trypsin are suitable for thermolysin except that the buffer should contain 1–5 mM calcium carbonate (which enhances thermostability to over 60°C) and a 2 h incubation is normally adequate at 37°C (1 h at 45°C). The enzyme is active in 6–8 M urea (with reduced thermostability) or 0.5% SDS, and is inhibited by EDTA. The pH optimum is between 7 and 8.

(iii) *Proline-specific endopeptidase (EC 3.4.21.76)*. A useful endopeptidase specific for Pro-X bonds (some Ala-X may also be cleaved slowly) has been demonstrated in a number of animal tissues. Although it has been used in a number of sequencing studies, the enzyme is not available commercially from these sources. An equivalent protease from *Flavobacterium meningoseptum* (49) can now be obtained from Seikagaku Kogyo Co. Ltd, Tokyo or Miles Laboratories. The enzyme does not work well on large proteins but gives good results when used to sub-digest large peptides (up to mol. wt 6000) using the following procedure.

(a) Incubate 0.1 μmol of peptide in 50 mM phosphate buffer pH 7.0 with 0.075 units (as defined by suppliers) of enzyme (~1% (w/w) enzyme: peptide) at 30°C for 2 h.
(b) Terminate the reaction and purify the peptides as above.

4.1.4 *Less specific proteases*

(i) *Elastase (EC 3.4.23.36)*. Elastase, a serine protease, has a rather wide specificity with preference for small non-polar residues such as alanine, although cleavage at serine, glycine, valine and leucine has been reported. Incubation conditions similar to those for trypsin and chymotrypsin [i.e. 100 mM NH_4HCO_3, at 25–37°C, 1–4 h with up to 3% (w/w) of enzyme] are suitable.

(ii) *Pepsin (EC 3.4.23.1)*. Pepsin shows particularly broad specificity and the most susceptible bonds vary widely from one protein to another. Although bonds involving phenylalanine and leucine are preferred, many others are also cleaved to some extent. Cleavage may occur on either side of a susceptible residue. Arginine, lysine, proline and, interestingly, isoleucine-containing bonds are not hydrolysed. Pepsin requires a strongly acidic milieu (pH 3) for maximal activity and for this reason is used where there is a need to preserve a disulphide bond (see Section 2.2). Under more alkaline conditions the disulphides can rearrange.

(a) Dissolve the protein in 99% formic acid at a concentration of up to 20 mg/ml.
(b) Dissolve pepsin in 1 mM HCl at a concentration of up to 0.05 mg/ml.
(c) Dilute 1 vol. of the protein solution with vols of pepsin (to give 2% enzyme), mix rapidly and incubate at 25°C for 1.5–2 h. It is possible to restrict cleavages

to the most susceptible sites by reducing the enzyme concentration to 0.1–1% and the digestion time to less than 1 h.

(d) The digestion can be terminated as with trypsin.

(iii) *Papain (EC 3.4.22.2) and subtilisin (EC 3.4.21.14).* These enzymes are examples of very non-specific proteases that may be used as a last resort, particularly in the fragmentation of large peptides which are otherwise too long for complete sequencing and in which there are no suitably specific cleavage points. Reaction conditions described for trypsin will work well for subtilisin, but papain as a thiol protease requires, in addition, 1 mM reducing agent (2-mercaptoethanol or DTT) and 0.1 mM EDTA. With papain, care must also be taken to exclude thiol reagents such as iodoacetate. Subtilisin is active in 1% SDS, papain in 8 M urea. Finally, *pronase*, in reality a mixture of proteases, can be used when very extensive degradation of the protein or peptide is required.

4.2 Chemical cleavage methods

Cleavage of peptide bonds using chemical rather than enzymic methods has been of enormous help in the elucidation of primary structure. Some are particularly invaluable as complements to proteases because efficient cleavage can be produced at bonds for which no specific enzyme has yet been found. An additional advantage is that many of these methods are highly specific for amino acids such as methionine or tryptophan which occur relatively rarely in proteins. They can be used, therefore, to generate relatively few large fragments of the protein thereby facilitating automatic sequencing and reducing the problems associated with fractionating complex mixtures of smaller peptides. Although a fairly large number of cleavage reactions have been described in the literature, some are not very convenient [e.g. cleavage at proline by sodium metal in liquid ammonia at −33°C under totally anhydrous conditions! (50)] and only a handful are in routine use. The most successful and reliable of these are described below.

4.2.1 *Asparagine–glycine bonds*

It is thought that when the next amino acid in the chain is glycine the asparaginyl side-chain can cyclize relatively easily to form a substituted succinimide. This species is liable to nucleophilic attack by hydroxylamine according to *Scheme 1* (51).

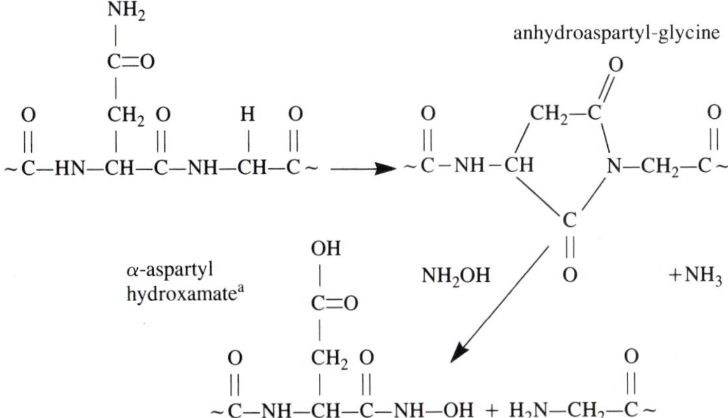

Scheme 1. Effects of hydroxylamine at pH 9.0 on Asp-Gly bonds. [a]β-Aspartyl hydroxamate also found.

A suitable method is that given in ref. 52.

(i) Dissolve the protein (reduced and carboxymethylated if possible) in 6 M guanidine–HCl, 2 M hydroxylamine-HCl to pH 9.0 with 4.5 M LiOH.
(ii) Incubate at 20–25°C for 3–4 h maintaining the pH at 9.0 with further additions of 4.5 M LiOH.

4.2.2 *Aspartic acid–X bonds* (also called partial acid hydrolysis).

Although not a very popular tool, this method (53) is a possible alternative to the use of *Staphylococcus aureus* V8 protease, particularly for proteins of limited solubility. The cleavage obtained is much more extensive than with the enzyme but rates vary depending on the nature of the X amino acid. Asp-Pro bonds appear to be much the most sensitive and may even cleave if acidic conditions are used for protein purification or during the procedure for cleavage at methionine with cyanogen bromide. Most stable bonds appear to be those involving hydrophobic residues such as leucine or valine. The conditions described below do not lead to extensive deamidation of amide side-chains except when they are at the C terminus of the protein or peptide.

(i) Dissolve the protein in dilute HCl (220 μl/100 ml) or dilute formic acid (2 ml/100 ml), pH 2 to 1–2 mg protein/ml. Alternatively, the protein can be dissolved in concentrated formic acid and then diluted.
(ii) Seal the tubes under vacuum and heat at 108°C for 2 h.
(iii) Open the tubes and dry under vacuum.

Other conditions that have been used include 2 HCl at 100°C for 30 min and 12 M HCl at 37°C for 3–7 days (54). Where specificity for the Asp-Pro bond is required, conditions such as 10% acetic acid adjusted to pH 2.5 with pyridine in 7 M guanidine–HCl for 24–96 h at 40°C (55) or 75% formic acid in 7 M guanidine–HCl for 48 h at 37°C may be successful (56). Note, however, that the yields and number of non-specific cleavages vary considerably depending on the particular protein.

It is worth pointing out that harsh acid treatment of proteins and peptides can cause side effects. These include cyclization of N-terminal glutamine residues to pyrrolidone carboxylic acid which renders the peptide refractory to Edman degradation, disulphide exchange and partial destruction of tryptophan.

4.2.3 *Cysteine*

Cysteine represents an attractive candidate for the targeting of cleavage methods because this residue is relatively rare in most proteins. Considerable attention has been paid to the development of efficient specific methods but so far with only limited success. The best of these utilizes conversion of the -SH moiety to the thiocyano group. This can cyclize to an acyliminothiazolidine followed by rapid hydrolysis to give the N-terminal peptide and a C-terminal peptide whose first residue contains the 2-iminothiazolidinyl function. Generation of the thiocyanate derivative is readily achieved with 2-nitro-5-thiocyanobenzoic acid (NTCB) which can be synthesized from 5,5′-dithionitrobenzoic acid (DTNB) using NaCN (57).

(i) Dissolve protein in 6 M guanidine–HCl, 0.2 M Tris–acetate pH 8.0 containing up to 1 mM DTT (if disulphides are present add 10 mM DTT and incubate at room temperature for 2 h).

(ii) Add a 5-fold molar excess (over *total* thiol) of NTCB, readjust pH to 8.0 with 1 M NaOH if necessary, and incubate at 37°C for 15–30 min.
(iii) Acidify the reaction mixture to pH 4.0 with acetic acid, cool to 4°C, remove the protein by dialysis or gel filtration and freeze dry.
(iv) For cleavage, dissolve the protein in 6 M guanidine–HCl, 0.1 M sodium borate pH 9.0 and incubate for 12–18 h at 37°C.

An alternative two-step procedure for S-cyanylation is also available. This involves treatment of the protein with DTNB followed by generation of the thiocyano derivative with KCN (58).

(i) Incubate the protein in 6 M guanidine–HCl, 0.2 M Tris–acetate pH 8.2 with 0.05 M DTNB at room temperature for up to 18 h.
(ii) Dialyse, add 5 mM KCN in the same buffer and continue incubation for up to 24 h. Then proceed to step (iii) above.

It is very important to appreciate that this process will generate peptides blocked at their N termini by the modified cysteine derivative (the iminothiazolidine). Without generation of a free amino group, therefore, the Edman degradation procedure will be inoperable. Considerable success with unblocking has been reported using Raney nickel (59).

4.2.4 *Histidine*

Histidine is another attractive candidate for cleavage, again because of its relative rarity in many proteins. Chemically it would also appear to be a distinctive candidate for attack. However, the only reported method with any success involves the use of *N*-bromosuccinimide (in pyridine–acetate pH 3.3, for 1 h at 100°C). The side reactions involving tryptophan, tyrosine and sulphur-containing residues as well as the potential lability at this temperature of sensitive peptide bonds make this an unattractive method.

4.2.5 *Proline*

Similarly, procedures for cleavage at this residue involving sodium hydrazide in hydrazine–ether at 0°C for up to 60 min (60) or lithium aluminium hydride in anhydrous tetrahydrofuran (61) produce additional non-specific degradation products.

4.2.6 *Methionine*

Methionine residues are rarely present at greater than 2–3 mole % and therefore offer an excellent target for chemical attack. Fortunately, a most effective method (62) has been devised for cleaving the polypeptide chain at such residues using cyanogen bromide (CNBr). The method is popular not only because of its specificity and high efficiency but because it generates a C-terminal residue, homoserine or homoserine lactone (see *Scheme 2*), which can subsequently be used to effect high-yield coupling to insoluble supports (see Chapters 3 and 6).

(i) Dissolve the protein (pre-treated with 0.1 M 2-mercaptoethanol or DTT at 37°C for 2 h) to about 10 mg/ml and add 100- to 1000-fold molar excess of CNBr over methionine either as a solid or as a solution (10–20 M) in acetonitrile (the

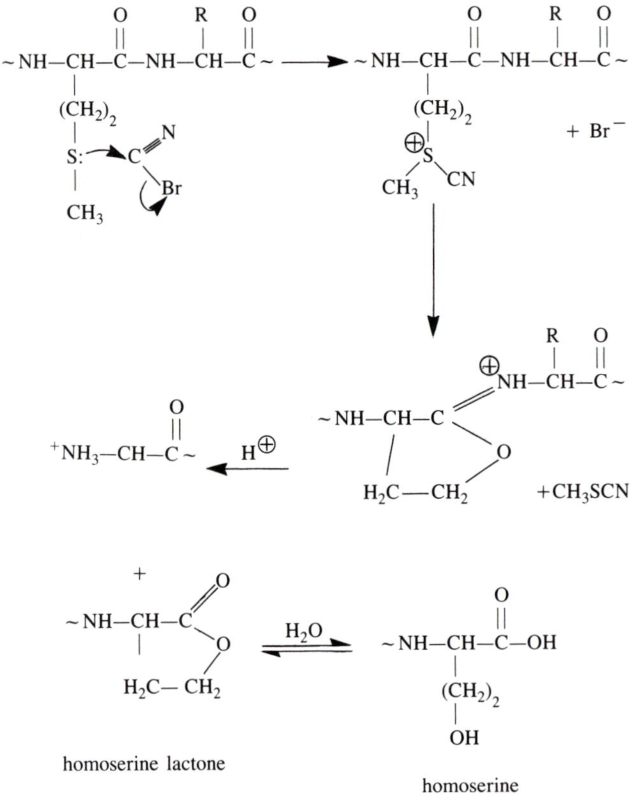

homoserine lactone

homoserine

Scheme 2.

higher the level of CNBr, the greater the chance of non-specific reactions). *Weigh out CNBr in a fume hood—biproducts are toxic.* Only colourless stocks of CNBr should be used.

(ii) Flush with N_2, seal and incubate in the dark at room temperature for 18–24 h.

(iii) Add 10 vol of water, rotary evaporate to remove excess CNBr, freeze and lyophilize.

Note: care should also be taken in emptying the cold trap of the freeze-drier. Allow the contents to melt in the presence of strong KOH to maintain an alkaline pH, add an equal volume of strong hypochlorite solution and wash away with copious amounts of water.

Under these conditions, CNBr will cleave most Met-X bonds in near quantitative yield, to leave peptides with a C-terminal homoserine residue. Some methionyl bonds may give poor yields, particularly Met-Thr and Met-Ser. The chemical reasons for this are illustrated in *Figure 5*. Cleavage of these bonds can be markedly improved by carrying out the cleavage in 70% TFA or by subsequent treatment of the freeze-dried digest, with 70% TFA (63). Occasionally cleavage at tyrosine and tryptophan residues and other non-specific cleavages may occur. Some, if not most of these, result from the strongly acidic conditions used.

Peptide bonds involving methionine sulphone or sulphoxide are *not* cleaved by CNBr.

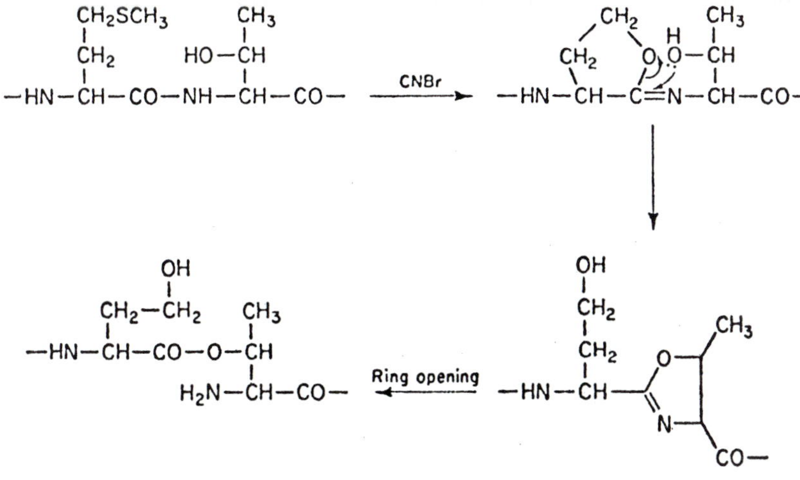

Figure 5. The N- to O-acyl shift during CNBr cleavage of Met-Ser or Met-Thr bonds.

Other reactions which can occur, particularly when high levels of CNBr are employed include oxidation of cysteine and cystine and cysteic acid and bromination of tyrosine side-chains. Limited cleavage at methionine has been achieved using a 50-fold molar excess of CNBr in 0.66 mM phosphate buffer pH 7.0 for 1 h at 20°C (64) but when such pH conditions are employed care must be taken not to effect modifications at other residues. Mild acid cleavage of Asp-Pro bonds is reduced at 4°C.

4.2.7 *Tryptophan*

The individual character of the indole side chain of tryptophan and relative rarity of this residue in most proteins has made this amino acid an attractive target for selective cleavage.

N-bromosuccinimide was probably the best known agent for the oxidation and cleavage of peptide chains at the carboxyl side of tryptophan. In practice the methods devised (65) are inefficient (5–50% cleavage), non-specific and can modify several amino acids, (e.g. histidine and tyrosine) without cleavage, producing products which cannot subsequently be recognized during sequencing. Methionine is oxidized and cysteine converted to cystine and cysteic acid. BNPS-skatole [2-(2-nitrophenylsulphonyl)-3-methyl-3-bromoindolenine] shows much greater specificity for tryptophan, and fewer side reactions have been reported. Yields of up to 70% using a 10-fold molar excess in 75% acetic acid at 37°C for 24 h can be obtained. Other methods for cleavage of tryptophan include incubation in dimethylsulphoxide and halogen acid (66), treatment with CNBr in heptafluorobutyric acid (67) and exposure to tribromocresol (68). None of these is as effective as BNPS-skatole.

However, in recent years *o*-iodosobenzoic acid (69) has become accepted as the most reliable of these tryptophan-directed reagents, with yields of 70–90% for most tryptophanyl bonds. The method is relatively simple and under the conditions described below only bonds containing tyrosine are also liable to attack. This can be largely prevented by the inclusion of excess tyrosine or of 1–20 mole % *p*-cresol (70) and by using reagents of the highest purity.

(i) Dissolve the protein in 80% acetic acid containing 4 M guanidine–HCl and 10–20 mole % *p*-cresol, to give a protein concentration of 5–10 mg/ml.
(ii) Flush with N_2, add 3 mg of *o*-iodosobenzoic acid/mg protein and incubate for 24 h at room temperature in the dark.
(iii) Terminate the reaction by dilution with 10 volumes of water and freeze-dry.
(iv) Alternatively, apply the digest directly to a column (200 × 1 cm) of bio-Gel P30 (Bio-Rad) equilibrated in 70% formic acid. The top third of the column should be protected from light by aluminium foil.

Tryptophan cleavage methods can be used to generate a C-terminal spirolactone and this can be very useful in solid-phase sequencing since the lactone couples very readily to amino-containing supports.

4.2.8 *Tyrosine*

In the previous section, comment was made that many of the agents used for cleavage to tryptophan residues also are effective but to a lesser extent, at attacking tyrosyl bonds. Traditionally *N*-bromosuccinimide which produces the dibromodienone spirolactone was in common use. Cleavage is performed at room temperature with 0.1 mg *N*-bromosuccinimide/mg of protein added three times over a period of 6–8 h to the protein dissolved in 50% acetic acid (71). The mixture can then be diluted and freeze-dried. Under these conditions cleavage yields of 30–80% can be expected but at the high reagent levels employed it should be appreciated that most of the tryptophan-containing peptide bonds are also liable to scission.

4.2.9 *N-terminal blocking*

Modification of the N-terminal amino group with formyl, acetyl or other acyl groupings or conversion to a pyroglutamyl residue is an increasingly frequent observation and it may be that up to 50% of all proteins are blocked in this way. Since such polypeptides are refractory to the Edman degradation chemistry, further procedures are required if sequence information is to be obtained. In general, simple chemical unblocking can occasionally be successful with formyl and pyroglutamyl residues but acetyl or other acyl moieties are much more difficult to remove. The method involves the incubation of protein with 0.1–3 M HCl in anhydrous methanol (72, 73). The time (up to 48 h) and temperature (20–37°C) of the incubation can be varied to effect maximal unblocking. The principal risk in this procedure is in cleavage of internal peptide bonds and this possibility should be carefully checked using, for example, SDS–PAGE. Depending on the identity of the acetylated residue, this method can also effect a degree of deacetylation. The more involved procedure of Schmer and Kreil (74) utilizing anhydrous hydrazine is not much more effective.

The pyrrolidone carboxylic acid residue may also be removed by incubation with the enzyme pyroglutamate aminopeptidase (EC 3.4.11.8) (75) in 0.1 M sodium phosphate pH 8.0 containing 5 mM DTT and 10 mM EDTA. The reaction may be allowed to proceed for up to 18 h at room temperature. The effectiveness of this approach varies considerably with the protein. No deacetylating biological activity has yet been discovered. Instead it seems that the acetyl-amino acid moiety is removed. The efficiency of the enzyme responsible varies with the amino acid but more importantly, peptides are much more amenable to degradation than intact proteins.

If free N termini are not generated by these methods, it will be necessary to digest the polypeptide into large fragments by limited proteolysis or into small peptides by more thorough proteolytic digestion. A variety of approaches can then be used, the most convenient and effective of which is mass spectrometry (see Section 3.1.4, Chapter 5).

5. REFERENCES

1. Soby,L.M. and Johnson,P. (1981) *Anal. Biochem.*, **113**, 149.
2. Swadesh,J.K., Thannhauser,T.W. and Scheraga,H.A. (1984) *Anal. Biochem.*, **141**, 397.
3. Liu,T.Y. and Chang,Y.H. (1971) *J. Biol. Chem.*, **216**, 2842.
4. Penke,B., Ferenczi,R. and Kovacs,K. (1974) *Anal. Biochem.*, **60**, 45.
5. Simpson,R.J., Neuberger,M.R. and Liu,T.Y. (1976) *J. Biol. Chem.*, **251**, 1936.
6. Capony,J.P. and Demaille,J.G. (1983) *Anal. Biochem.*, **141**, 397.
7. Morrice,N.M. and Aitken,A. (1985) *Anal. Biochem.*, **148**, 207.
8. Sefton,B.M., Hunter,T., Ball,E.H. and Singer,S.J. (1981) *Cell*, **24**, 165.
9. Martensen,T.M. (1982) *J. Biol. Chem.*, **257**, 9648.
10. Yang,J.C., Fujitaki,J.M. and Smith,R.A. (1982) *Anal. Biochem.*, **122**, 360.
11. Westall,F. and Hesser,W. (1974) *Anal. Biochem.*, **61**, 610.
12. Fleury,M.O. and Ashley,D.V. (1983) *Anal. Biochem.*, **133**, 330.
13. Cooper,J.D.H., Lewis,M.T. and Turnell,D.C. (1984) *J. Chromatogr.*, **285**, 484.
14. Schuster,R. and Apfel,A. (1986) HPLC Application Note, Hewlett-Packard Bioscience. A new technique for the analysis of primary and secondary amino acids.
15. Hunkapiller,M.W., Strickler,J.E. and Wilson,K.J. (1984) *Science*, **226**, 304.
16. Moore,S. and Stein,W.H. (1963) In *Methods in Enzymology*, Colowick,S.P. and Kaplan,N.O. (eds), Academic Press, New York, Vol. 6, p. 819.
17. Cavins,J.F. and Friedman,M. (1970) *Anal. Biochem.*, **35**, 489.
18. Lottspeich,F. (1985) *J. Chromatogr.*, **326**, 321.
19. Heinrikson,R.L. and Meredith,S.C. (1983) *Anal. Biochem.*, **136**, 65.
20. Bidlingmeyer,B.A., Cohen,S.A. and Tarvin,T.L. (1984) *J. Chromatogr.*, **336**, 93.
21. Plapp,B.V., Moore,S. and Stein,W.H. (1971) *J. Biol. Chem.*, **246**, 939.
22. Yamada,S. and Itano,H.A. (1966) *Biochim. Biophys. Acta*, **130**, 538.
23. Brown,J.R. and Hartley,B.S. (1963) *Biochem. J.*, **89**, 59.
24. Brown,J.R. and Hartley,B.S. (1966) *Biochem. J.*, **101**, 214.
25. Meek,J.L. (1980) *Proc. Natl. Acad. Sci. USA.*, **77**, 1632.
26. Browne,C.A., Bennett,H.P.J. and Solomon,S. (1982) *Anal. Biochem.*, **124**, 201.
27. Walsh,K.A. and Sasagawa,T. (1984) In *Methods in Enzymology*, Wold,F. and Moldave,K. (eds), Academic Press, New York, Vol. 106, p. 22.
28. Biscoglio, Jimenez Bonino,M.J., Fukushima,J.G. and Cascone,O. (1986) *Anal. Biochem.*, **157**, 8.
29. Sasagawa,T., Titani,K. and Walsh,K.A. (1983) *Anal. Biochem.*, **128**, 371.
30. Sasagawa,T., Titani,K. and Walsh,K.A. (1983) *Anal. Biochem.*, **134**, 224.
31. Krebs,E.G. (1986) In *The Enzymes*. Boyer,P.D. and Krebs,E.G. (eds), Academic Press, Orlando, 3rd Edition, Vol. 17a, p. 3.
32. Meyer,H.E., Hoffman-Posorske,E., Korte,H. and Heilmeyer,L.M.C. (1986) *FEBS Lett.*, **204**, 61.
33. Holmes,C.F.B. (1987) *FEBS Lett.*, **215**, 21.
34. Barber,M., Bordoli,R.S., Sedgwick,R.D., Tyler,A.N. (1981) *Nature*, **293**, 270.
35. Holmes,C.F.B., Tonks,N.K., Major,H. and Cohen,P. (1987) *Biochim. Biophys. Acta*, **929**, 208.
36. Drapeau,G.R., Boily,Y. and Houmard,J. (1972) *J. Biol. Chem.*, **247**, 6720.
37. Drapeau,G.R. (1976) In *Methods in Enzymology*. Lorland,L. (ed.), Academic Press, New York, Vol. 45, p. 469.
38. Findlay,J.B.C., Pappin,D.J.C. and Brett,M. (1981) *Nature*, **293**, 314.
39. Noreau,J. and Drapeau,G.R. (1979) *J. Bacteriol.*, **140**, 911.
40. Patthey,L. and Smith,E.L. (1975) *J. Biol. Chem.*, **250**, 557.
41. Yamasaki,R.B., Vega,A. and Feeney,R.E. (1980) *Anal. Biochem.*, **109**, 32.

42. Mitchell,W.M. and Harrington,W.F. (1968) *J. Biol. Chem.*, **243**, 4683.
43. Schenkein,I., Levy,M., Franklin,E.C. and Frangione,B. (1977) *Arch. Biochem. Biophys.*, **182**, 64.
44. Mitchell,W.M. (1977) In *Methods in Enzymology*. Hirs,C.H.W. and Timasheff,S.N. (eds), Academic Press, New York, Vol. 47, p. 165.
45. Doonan,S., Doonan,H.J., Hanford,R., Vernon,C.A., Walker,J.M., Airoldi,P. da S., Bossa,F., Barra,D., Carloni,M., Fasella,P. and Riva,F. (1975) *Biochem. J.*, **149**, 497.
46. Doonan,S. and Fahmy,H.M.A. (1975) *Eur. J. Biochem.*, **56**, 421.
47. See Boehringer Mannheim catalogue.
48. Heinrikson,R.L. (1977) In *Methods in Enzymology*. Hirs,C.H.W. and Timasheff,S.N. (eds), Academic Press, New York, Vol. 47, p. 175.
49. Yoshimoto,T., Walter,R. and Tsuru,D. (1980) *J. Biol. Chem.*, **255**, 4786.
50. Hempel,J. and Jornvall,H. (1985) *Anal. Biochem.*, **151**, 255.
51. Bornstein,P. and Balian,G. (1977) In *Methods in Enzymology*. Hirs,C.H.W. and Timasheff,S.N. (eds), Academic Press, New York, Vol. 47, p. 132.
52. Enfield,D.L., Ericsson,L.H., Fujikawa,K., Walsh,K.A., Neurath,H. and Titani,R. (1980) *Biochemistry*, **19**, 659.
53. Inglis,A.S., McKern,N.M. and Stike,P.M. (1979) *Proc. Aust. Biochem. Soc.*, **12**, 12.
54. Milstein,C. and Sanger,F. (1961) *Biochem. J.*, **79**, 456.
55. Fraser,K.J., Poulson,K. and Harber,E. (1972) *Biochemistry*, **11**, 4974.
56. Jauregui-Adell,J. and Marti,J. (1975) *Anal. Biochem.*, **69**, 468.
57. Jacobson,G.R., Schaffer,M.H., Stark,G.R. and Vanaman,T.C. (1973) *J. Biol. Chem.*, **248**, 6583.
58. Stark,G.R. (1977) In *Methods in Enzymology*. Hirs,C.H.W. and Timasheff,S.N. (eds), Academic Press, New York, Vol. 47, p. 124.
59. Otieno,S. (1978) *Biochemistry*, **17**, 5468.
60. Kauffmana,T. and Sobel,J. (1966) *Justins Liebig's Ann. Chem.*, **698**, 235.
61. Ruttenberg,M.A., King,T.P. and Craig,L.C. (1965) *Biochemistry*, **4**, 11.
62. Gross,E. and Witkop,B. (1961) *J. Am. Chem. Soc.*, **83**, 1510.
63. Pappin,D.J.C. and Findlay,J.B.C. (1984) *Biochem. J.*, **217**, 605.
64. Pellicone,C., Nullans,G. and Virmaux,N. (1985) *FEBS Lett.*, **181**, 179.
65. Fontana,A. and Toniolo,C. (1976) *Prog. Chem. Org. Natural Compounds*, **33**, 309.
66. Savige,W.E. and Fontana,A. (1977) In *Methods in Enzymology*. Hirs,C.H.W. and Timasheff,S.N. (eds), Academic Press, New York, Vol. 47, p. 459.
67. Ozols,J. and Gerard,C. (1977) *J. Biol. Chem.*, **252**, 8549.
68. Burstein,Y. and Patchornik,A. (1972) *Biochemistry*, **11**, 4641.
69. Mohoney,W.C., Smith,P.K. and Hermondson,M.A. (1981) *Biochemistry*, **20**, 443.
70. Fontana,A., Dalzoppo,D., Grandi,C. and Zambonin,M. (1981) *Biochemistry*, **20**, 6697.
71. Hurley,C.K. and Stout,J.T. (1980) *Biochemistry*, **19**, 410.
72. Sheehan,J.C. and Yang,D.-D.H. (1958) *J. Am. Chem. Soc.*, **80**, 1154.
73. Kawasaki,I. and Itano,H.A. (1972) *Anal. Biochem.*, **48**, 546.
74. Schmer,G. and Kreil,G. (1969) *Anal. Biochem.*, **29**, 186.
75. Podell,D.N. and Abraham,G.N. (1978) *Biochem. Biophys. Res. Commun.*, **81**, 176.

CHAPTER 3

Automated solid-phase microsequencing

J.B.C.FINDLAY, D.J.C.PAPPIN and J.N.KEEN

1. INTRODUCTION

Automation of the Edman degradation sequencing chemistry centres around the delivery of reagents under the most appropriate conditions and subsequent removal of the reaction products in such a way as to allow recovery of the released amino acid derivative whilst leaving the remainder of the protein/peptide behind in the reaction chamber. Historically two approaches have evolved. The spinning cup technology relied on differential solubility in the washing solutions of protein/peptide versus derivatized amino acid and on the volatility of certain of the reagents and biproducts. The solid-phase approach pioneered by Laursen (1) was in many ways simpler and more attractive, for by covalently attaching the protein or peptide to an insoluble support, the washing and recovery processes could be rendered much more straightforward and efficient.

It was perhaps a surprise, therefore, that the first of these technologies was initially the more successful due, in some measure, to the development of efficient procedures and equipment. However, the solid-phase approach also suffered one serious drawback namely that the coupling of peptide/protein to insoluble support was widely regarded as inefficient. In some ways this impression was justified for the chemistries of the coupling procedures were neither universally applicable nor particularly well developed. Added to this, the characterization of many of the supports used was inadequate, leading to unexplained inefficiency in the coupling and sequencing processes.

Since that time two developments have taken place; the spinning cup approach has been further refined to reappear in the guise of the gas-phase sequencer (see Chapter 4) whilst the solid-phase chemistries have been standardized and made much more efficient. Although they may now appear as competing technologies, this is unlikely to remain so indefinitely for as new chemical procedures are invented, a single strategy involving covalent attachment may well emerge.

At this point, it is perhaps worthwhile summarizing briefly the advantages and disadvantages of the solid-phase approach.

1.1 **Advantages**

(i) Initial and repetitive yields can be very high, leading to very long sequencing runs (>100 residues identified from 2−3 nmol of protein).

(ii) Hydrophobic peptides are not removed by the various reactants and solvents.

(iii) The washing procedures can be made very efficient, resulting in low chemical noise.

(iv) Charged derivatives of the amino acids, for example phosphorylated residues, are not retained in the reaction chamber and can be efficiently recovered.

(v) Cycle times are much reduced compared with other technologies.
(vi) Material can be solubilized efficiently and coupled to solid-phase supports in detergents such as sodium dodecylsulphate (SDS).

1.2 **Disadvantages**

Coupling procedures remain the greatest problem but methods are slowly improving to the point where carboxyl coupling may become routine. In methods used for attachment via carboxyl- or amino-containing side-chains, the recovery of these residues during sequencing is variable and may hinder identification.

2. SAMPLE PREPARATION

It goes without saying that proper attention to sample purity is an essential prerequisite to good sequencing results. Chapter 1 in this volume or a companion volume on protein isolation (2) contain detailed procedures for sample purification. It is worthwhile, however, summarizing a few important points.

Small molecules are just as much a problem as contaminating peptides, particularly if they contain primary amines which interfere with the coupling and sequencing procedures. Perhaps the most common source of problems is ammonium-containing salts, particularly the often used bicarbonate. Lyophilization is often thought to completely remove this salt but in our experience this is never achieved without some form of de-salting. Other commonly used substances which should be avoided or removed include Tris, pyridine, glycine, bicine, amino sugars, polybuffers, ampholytes and some detergents. Although solid-phase sequencing will still occur in the presence of most of these, they do often harbour unsuspected impurities which seriously reduce sequencing efficiency. Phospholipids, carbohydrates and nucleic acids each present their own particular difficulties and should be removed.

Methods used for purification include dialysis or gel filtration followed by freeze-drying, adsorption and elution (with acetonitrile or water) using Sep-Pak™ cartridges, high-performance liquid chromatography (particularly reverse-phase), hydrophobic interaction chromatography and precipitation with 10–20 vols of cold acetone, 5–10 vols of cold diethyl ether or ether:ethanol combinations (1:1 to 3:1 v/v)—good for detergent removal.

Finally, SDS–polyacrylamide gel electrophoresis (SDS–PAGE) followed by electroelution presents a rapid and very useful way of preparing polypeptides for sequencing. A simple procedure is given in *Table 1*. We now routinely use this method for sample preparation. It can be conveniently combined with limited proteolysis in 0.1% SDS using *Staphylococcus aureus* V8 protease, endoproteinase Lys-C or Arg-C or a number of other proteolytic enzymes (see Chapter 6, ref. 3). In this way a number of large fragments can be generated, coupled to resins and subjected to long sequence runs thereby giving very substantial amounts of sequence data. This type of approach is especially useful for integral membrane proteins (3), for which the solid-phase technology is particularly well-suited.

Table 1. Electroelution of proteins from SDS–PAGE.

Equipment and reagents

1. Flat-bed gel tank of the type in common use for DNA separations or adapted tube gel apparatus.
2. Dialysis tubing—boiled for 15 min in 0.1 M sodium carbonate with 20 mM EDTA, then thoroughly rinsed with distilled water—use Spectrapor for small proteins/peptides.
3. Coomassie stain solution: 0.1% (w/v) Coomassie brilliant blue in 50% (v/v) aqueous methanol, 7% (v/v) acetic acid. The de-stain solution is identical, minus dye.

Procedure

1. Stain the gel very briefly with Coomassie blue (5–10 min) then rinse in de-stain solution until the protein bands become visible (10–15 min). Keep the stain and de-stain times as short as possible for maximum yields.
2. Excise the band of interest. Do *not* break up or homogenize the gel slice.
3. Fill the flat-bed tank with buffer (25 mM Tris–Gly pH 8.5, 0.1% w/v SDS), 50 mM Tris–acetate pH 7.8, 0.1% SDS, 0.1M sodium bicarbonate pH 7.8, or 0.1 M sodium phosphate pH 7.8, 0.1% SDS) to a level approximately 1 cm above the platform. 0.1% (v/v) 2-mercaptoethanol or 2–5 mM DTT can be added to the buffer if required. For membrane proteins, the SDS concentration of the buffer can be raised to 1% (w/v) to ensure solubility.
4. Cut a length of dialysis tubing long enough to take the gel slice plus ~2 cm at either end. Clip one end with a Mediclip and fill the tubing with buffer from the tank. Place the excised gel slice in the tubing and gently squeeze out most of the liquid before sealing. Position the gel slice on one side of the dialysis bag. Care should be taken to avoid trapping any air bubbles.
5. Place the dialysis bag onto the platform of the electrophoresis tank. The level of buffer should be adjusted so that it just covers the bag.
6. Electrophorese for 3–20 h at 25–100 V constant current (20–150 mA) then reverse the current for ~30 sec (in order to electrophorese the protein off the dialysis membrane surface). Longer staining/de-staining periods require more extended electroelution times.
7. Remove the gel slice from the bag. Re-stain to check that the protein has been eluted. Any small fragments of gel that remain *must* be removed. If necessary, decant the solution and centrifuge briefly or filter to remove any small pieces. Re-seal the bag and dialyse the protein solution against at least five changes of distilled water (5 litres each) over 2–3 days at 0–4°C. The dialysis solutions can contain up to 0.25% (w/v) SDS to keep the protein in solution, but make the last change against distilled water only. The Coomassie dye stays with the protein throughout and can thus be used as an indicator of the progress of the electroelution.

Modifications

The procedure can be used with all forms of staining (except for silver staining which blocks the protein) and with any apparatus based on tube gels where the dialysis bag is fixed onto the bottom of the tube. To eliminate potentially the greatest loss-producing step, the proteins can be stained *during* electrophoresis by adding Coomassie dye (25 mg/l) to the cathode buffer. No fixing and staining is then required.

3. METHODOLOGY

3.1 **Preparation of supports**

Supports need to possess the properties of stability to sequencing chemistries, resistance to mechanical deformation, high porosity and large surface area/unit mass. So far most use has been made of the so-called ‘soft’ resins, generally derived from cross-linked polystyrenes, or more successfully, of porous glass beads which resist compaction by pressure and exhibit good flow properties.

3.1.1 *Polystyrenes*

Tables 2 and *3* present methods for the preparation of amino-polystyrene (4) and

Table 2. Preparation of amino-polystyrene.

Carry out the procedure in a fume hood.

1. Wash polystyrene beads (Biobeads SX-1, 400 mesh) extensively with high purity benzene, chloroform and finally methanol. Dry under vacuum.
2. Add 200 g of the washed beads slowly and with gentle stirring to 250 ml of 95% HNO_3 pre-cooled to near 0°C. The temperature should not be allowed to rise above 2–3°C.
3. Incubate with gentle stirring for a further 1 h at 0°C.
4. Pour onto crushed ice in a sintered glass funnel. Wash thoroughly with dioxan, water and finally methanol and dry the resultant nitropolystyrene under vacuum.
5. Add the nitropolystyrene with gentle stirring to 400 ml of dimethylformamide (DMF), maintaining the temperature at 75°C. (DMF should be freshly purified over Al_2O_3.)
6. Add slowly and very carefully (exothermic reaction!) a warm solution of 200 g of $SnCl_2 \cdot 2H_2O$ in 150 ml of DMF.
7. Increase the temperature to 140–150°C and incubate at this temperature for about 15 min.
8. Cool, add 150 ml of concentrated HCl and incubate at 100°C for 1 h.
9. Cool and thoroughly wash the resin on a sintered glass funnel with water, 0.1 M HCl, 50% aqueous pyridine and triethylamine:DMF (1:3 v/v). Check that all the Cl^- ions are removed using silver nitrate acidified with HNO_3 (chloride will give a white precipitate).
10. Wash thoroughly with water followed by methanol; dry overnight under vacuum and store at room temperature (final yield should be >80% w/w).

Structure: 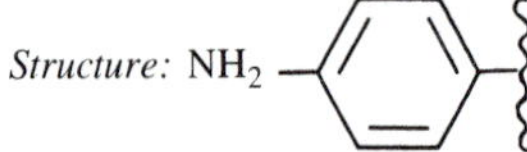

Table 3. Preparation of 'capped' TETA-polystyrene.

1. Add 10 ml of tetrahydrofuran (THF) and 10 ml of triethylenetetramine (TETA) to 1 g of chloromethylated polystyrene (200–400 mesh, 4.3 mEquiv/g from Bio-Rad, Pierce).
2. Seal under N_2 and stir gently at 70°C for 5 h.
3. To react out excess chloromethyl groups, add 20 ml of ethanolamine, seal under N_2 and continue stirring gently at 70°C for a further 2 h.
4. Filter the resin and wash with 50 ml of methanol.
5. Add 20 ml of triethylamine and stir gently at 20°C for 15–20 min.
6. Filter the resin and wash with 100 ml of methanol, 100 ml of water and finally 100 ml of methanol.
7. Dry down *in vacuo* and store in sealed vials at −20°C for no more than 2 months.
8. Amine substitution should be of the order of 2.5 mmol/g resin.

Structure: 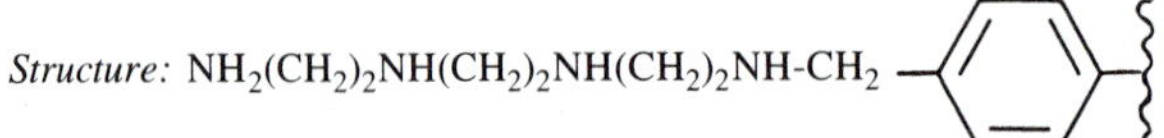

triethylenetetramine (TETA)-polystyrene (5). The efficiency of the latter resin can be severely compromised by failure to block (i.e. 'cap') the excess chloromethyl substituents which will subsequently react irreversibly with free amino groups (D.J.C.Pappin and J.B.C.Findlay, unpublished). As a consequence initial sequencing yields were often very low. In general, amino-polystyrene is most useful for smaller peptides with C-terminal lysine residues whilst TETA-polystyrene can be used for larger peptides terminating with or containing lysine. TETA-polystyrene was also particularly successful for peptides containing a homoserine lactone substituent on their C termini, generated as a result of cleavage of the protein with cyanogen bromide (CNBr). Accessibility to internal resin surfaces by polypeptides limited the usefulness of these polystyrene-based resins with proteins.

3.1.2 *Glasses*

Controlled pore glass beads have been the most successful support to date due largely to their rigidity, stability and relative ease of chemistry. They are by no means absolutely inert surfaces, however, and their chemical stability is not total. Pore-size is an important consideration. At first 75 Å glasses were used extensively but whilst this was generally satisfactory for peptides, penetration problems can be encountered with proteins. Mainly for that reason, pore sizes of 170–200 Å are more universally suitable, although it is possible to go up to 500 Å. These glasses should be carefully washed prior to derivatization. A simple protocol is given below.

(i) Boil gently for 10 min in 10 mM HCl.
(ii) Wash thoroughly with distilled water.
(iii) Wash thoroughly with anhydrous methanol.
(iv) Dry for not longer than 2 h at 200°C.
(v) Cool *in vacuo* over P_2O_5.

This glass may now be modified as detailed in *Tables 4* and *5* to give a number of very useful supports. Two further derivatized glasses which may have certain specialized applications are iodo-glass and amino-aryl glass. The first can be prepared by suspending aminopropyl glass (600 mg) in 4 ml of ethyl acetate containing 90 mg of re-crystallized iodoacetic acid and 100 mg of dicyclohexylcarbodiimide (8). The support is then washed extensively in water and acetone prior to drying *in vacuo* and storage at −20°C.

The amino-aryl support is produced by reaction of 3-aminopropyl glass (APG) with *p*-nitrobenzoyl chloride followed by reduction of the amine with sodium dithionite (9).

Table 4. Preparation of 3-aminopropyl and *N*-(2-aminoethyl)3-aminopropyl glasses (APG and AEAPG)[a].

1. Suspend 1–10 g of dry controlled pore glass (CPG 170 Å pore-size, 200–400 mesh, Sigma Chemical Corp) in a solution of 1% v/v 3-aminopropyltriethoxysilane (APG) or *N*-(2-aminoethyl)-3-aminopropyltrimethoxysilane[b] (AEAPG) [both reagents from Pierce]—in dry toluene[c]. Use sufficient silane solution to just cover the glass beads and sonicate briefly (or apply a light vacuum) to thoroughly de-gas the suspension.
2. Incubate for 15–18 h at 50°C.
3. Wash several times with methanol (HPLC grade) on a sintered glass filter.
4. Dry *in vacuo* and store desiccated at −20°C. Glass should be as white as the CPG starting material and is stable for at least 6 months.

Structures: $H_2N(CH_2)_3\,Si(-O-)_3$ CPG (APG)

$H_2N(CH_2)_2HN(CH_2)_3Si(-O-)_3$ CPG (AEAPG)

[a]The procedure is derived from ref. 6.
[b]Amine loadings are typically 200–250 moles amine/g glass and can be measured using the picric acid assay (*Table 6*).
[c]Dry toluene is prepared by distilling from sodium metal and benzophenone under argon.
[d]Thiol glasses can be prepared in much the same way using the appropriate silane reagent (e.g. 3-mercaptopropyltrimethoxysilane, Fluka).

Table 5. Preparation of DITC-glass[a].

1.	Suspend 5 g of 3-aminopropyl glass in 15 ml of a solution of dry tetrahydrofuran (THF)[b] containing 0.2 M *p*-phenylenediisothiocyanate (Fluka or Eastman Kodak, freshly crystallized from acetonitrile or acetone). Sonicate or briefly apply a light vacuum to de-gas the suspension.
2.	Seal under N_2 or argon and incubate for 2 h at room temperature with occasional gentle agitation.
3.	Wash several times with benzene (Fluka, Sequencer grade) then methanol (HPLC grade) on a sintered glass filter.
4.	Dry *in vacuo* and store desiccated at −20°C for up to 6 months. The glass should remain as white as the original starting material. Yellow colouration probably indicates some degree of polymerization although a very pale colour does not affect usefulness of the support[c].

[a]The procedure is derived from ref. 7.
[b]Dry THF is prepared by distilling from sodium metal and benzophenone under argon.
[c]Usually 85−95% of surface amines react with the isothiocyanate. Residual amines can be determined using the picric acid assay.

Table 6. Picric acid assay for amino groups on solid supports.

Reagents

(A) Picric acid	Add 300 mg of wet picric acid (Aldrich) to 200 ml of dichloromethane and dry over anhydrous Na_2SO_4 to remove any water. Filter the resulting solution and store in a tightly sealed light-tight bottle over 20 g of Na_2SO_4. (Stable at room temperature for up to 3 months.)
(B) Dichloromethane	Store in a light-tight container over anhydrous $NaHCO_3$ (200 ml/10 g).
(C) Triethylamine	4% in dichloromethane prepared immediately before use.

Protocol

1. Weigh accurately 5−10 mg of resin into a sintered glass funnel and wash with 2 ml of the picric acid solution (A), followed by 4 × 2 ml of dichloromethane (B).
2. Elute the bound picric acid into a volumetric flask (25 ml) using triethylamine in dichloromethane (C) and fill to the mark.
3. Read the absorbance at 358 nm against solution C and calculate the amount of amino groups/g resin (mmol/g) using

$$\frac{\text{Dilution (ml)} \times A_{308}}{\text{weight of resin (g)} \times E_{358}{}^{a} \times \text{path length}}$$

[a]The E_{358} value is 14 500 ml/cm/mmol.

The degree of amino group substitution can be assessed using the picric acid assay (*Table 6*).

All these procedures are optimized for glass beads but can equally well be applied to glass sheets (10).

3.2 Coupling procedures

The covalent attachment of peptides or proteins to insoluble supports is a critical step in the solid-phase approach to microsequencing. Although unfairly maligned in the past, very efficient methods are now available for a number of different chemical groups. New methods are also under development which promise to extend significantly the range of this sequencing strategy.

Table 7. Coupling to AP or AEAP supports.

1.	Dissolve the peptide in up to 100 μl of 50% aqueous *N*-methyl morpholine adjusted to pH 9.0 with TFA.
2.	Add 100 μl of DITC in DMF (10 mg/ml) and incubate under N_2 for 30 min at 45°C.
3.	Add the activated peptide solution to amino-resin (up to 40 mg) pre-swollen in 200 μl of DMF and incubate under N_2 for 45 min at 45°C.
4.	Block the excess amino groups with 100 μl of methylisothiocyanate in acetonitrile (1:1 v/v) for 45 min at 45°C.
5.	Wash the resin in a sintered glass funnel with several volumes of methanol and dry *in vacuo*.

Table 8. Coupling procedure for DITC-glass.

1.	Dissolve the lyophilized peptide (up to 2 nmol) in 50 μl of 0.2 M Na_2HPO_4/0.25% (w/v) SDS (pH 8–8.5). The SDS concentration may be increased to 2.0% (w/v) to improve solubility.
2.	Add this solution to 10 mg of DITC-glass (170 Å pore size, 200–400 mesh), prepared as in *Table 5*. Run the peptide solution over the surface of the glass to avoid trapping air and allow even coupling to the whole of the glass surface. Sonicate briefly and gently to remove trapped air bubbles and de-gas the suspension.
3.	Incubate the glass for 60 min at 56°C under nitrogen with occasional gentle agitation.
4.	Wash the glass with water and then methanol, both made slightly alkaline with 0.5% (v/v) *N*-propylamine (Fluka puriss) to remove non-covalently bound material. This procedure can be carried out in a small sintered glass funnel or bench centrifuge.
5.	Dry under vacuum. The coupled peptide can be stored dry for many months at −20°C.

This procedure applies to microsequencing but can be scaled up as appropriate for larger reaction chambers. It is possible to couple peptides to DITC-glass under non-aqueous conditions. First treat the dry peptide with 100 μl of anhydrous TFA for 30 min at room temperature. Dry down thoroughly *in vacuo* over KOH and re-dissolve in DMF instead of Na_2HPO_4/SDS as above. Add to DITC-glass as in step 2 and introduce 10 μl of triethylamine. Incubate for 60 min at 45°C under N_2 and then wash as above, substituting DMF for water.

3.2.1 *Lysine*

Lysine side-chains couple efficiently to amino-containing resins or glasses following modification of the amino group with *p*-phenylene diisothiocyanate (DITC), see *Table 7* (11). A better and more convenient approach involves direct coupling of the peptide/protein to DITC-activated glass, see *Table 8* (7). This latter method has had considerable success and for the vast majority of proteins and peptides, which usually contain lysine residues, is the method of choice. The procedure is generally not affected by most salts, detergents and dyes (e.g. Coomassie blue). Indeed the presence of up to 1% SDS can be a great advantage. However since amino-containing compounds will compete with the protein for binding, these impurities must be rigorously excluded from the sample. Although the N-terminus and all lysyl side-chains couple efficiently, usually enough of these residues remain unreacted to allow positive identification.

3.2.2 *Lactones (methionine and tryptophan)*

Cleavage of proteins with CNBr or *o*-iodosobenzoic acid generates derivatives of methionine and tryptophan at the C terminus of the resultant peptides. These can be readily converted to the corresponding homoserine lactone and spirolactone. Such

Table 9. Homoserine lactone coupling.

1.	Lactonize dried peptide for 60 min at room temperature in 0.1 ml of anhydrous TFA.
2.	Remove TFA over NaOH *in vacuo*.
3.	Dissolve the peptide in 0.05 ml of freshly purified DMF. For particularly hydrophobic material DMF–propan-2-ol (3:1 v/v) can be used.
4.	Add to 20 mg of aminopropyl-glass.
5.	Wash the tube with a further 20 μl of DMF (or alternative as above).
6.	Incubate for 20 min, at 56°C. Sonicate occasionally (and gently).
7.	Add 20 μl of re-distilled triethylamine.
8.	Incubate for 90 min, at 56°C. Sonicate occasionally.
9.	Wash the glass with methanol (2 × 2 ml), TFA (2 ml) and methanol (2 × 2 ml). Dry *in vacuo*.

Peptides may need to be de-salted prior to coupling.

Table 10. Coupling using carbodiimide.

1.	Dissolve peptide (1 nmol) in 100 μl of re-distilled pyridine–HCl pH 5.0.
2.	Add to washed and dried amino-aryl or amino-alkyl (the former is preferable) glass (20 mg).
3.	Add 1 mg of *N*-ethyl-(*N*-dimethylaminopropyl)carbodiimide–HCl (EDC) dissolved in 50 μl of water.
4.	Incubate for no more than 30 min at room temperature, with occasional gentle agitation.
5.	Wash thoroughly in water (2 × 1 ml), methanol (2 × 2 ml), 250 μl of anhydrous TFA, and finally methanol.
6.	Dry under vacuum and store at –20°C until required.

Where amino-alkyl resins are to be used, the peptides are usually dissolved in DMF (containing 1–2 μl of *N*-methylmorpholine if required) and dicyclohexylcarbodiimide added in anhydrous DMF. Incubate for up to 2 h at 40°C. Coupling efficiency can be increased by adding hydroxysuccinimide or hydroxybenzotriazole with the carbodiimide (2:1 v/v).
Samples should be free of acidic solvents, salts and detergents such as SDS.

chemical species readily and spontaneously react with amino groups on the resins/glasses (5,12). Provided the peptides/proteins are soluble in the near anhydrous conditions used for the reaction, coupling efficiencies can be as high as 90% (see *Table 9*). It is also possible to react the lactones with a diamine under the same conditions as used for the supports. The resultant product, after dialysis or filtration to remove excess amine, can then be coupled to DITC-glass.

3.2.3 *Carboxyl groups*

By far the most attractive candidate for attachment would be the C-terminal carboxyl group since a single method would be applicable to virtually all peptides and proteins. Unfortunately, current coupling methods involving carboxyl groups have efficiencies in the 20–50% range and it is very difficult to prevent side-chain reactions. However, incomplete reaction of the side-chain carboxyls has a hidden bonus for it allows identification of those residues which would otherwise be very difficult to assign if attachment was highly efficient.

Two methods have been used. The most popular utilizes water-soluble carbodiimides (*Table 10*), with or without temporary blocking at the amino-terminal with *t*-butoxycarbonyl azide (12,13). Side-chain carboxyls are thought to react less readily because, on reaction with the carbodiimide, they can rearrange from the O-acyl to the unreactive N-acyl derivative. The C-terminal carboxyl on the other hand forms the azolactone

Table 11. TFA anhydride coupling.

1.	Dissolve the peptide in 0.1 ml of anhydrous TFA and add 10 μl of TFA anhydride.
2.	Incubate for 60 min at room temperature under an inert atmosphere. Dry *in vacuo*.
3.	Dissolve the peptide in 50 μl of freshly purified dry DMF and add to 20 mg of APG.
4.	Incubate for 20 min, at 56°C, gently sonicate occasionally.
5.	Add 20 μl of re-distilled triethylamine.
6.	Incubate for 90 min at 56°C. Sonicate gently from time to time.
7.	Wash the glass with methanol (2 × 2 ml), TFA (2 ml) and methanol (2 × 2 ml). Dry *in vacuo*.

which has good reactivity with amino groups. A most important consideration is that the amino-aryl glasses may react much better than the amino-alkyl resins such as APG.

The second method (14,15) utilizes formation of a cyclic or mixed acid anhydride using trifluoroacetic acid (TFA) anhydride in TFA (*Table 11*). The chemistry of the process has not yet been confirmed but average coupling yields in the region of 30% can be achieved. The method is particularly attractive when used with peptides containing a C-terminal glutamic acid generated as a result of cleavage with *S.aureus* V8 protease (also now called endoproteinase Glu-C).

3.2.4 *Thiol groups*

Although not prevalent in proteins or peptides, there are occasions where covalent attachment via a thiol group (i.e. cysteine) can be a very positive advantage, not only from a sequencing perspective but also as a means of achieving peptide purity. More efficient coupling is achieved with the iodo-glasses rather than relying on disulphide bond formation with thiol-resins. A simple method is given below.

(i) Dissolve *reduced* peptide in 50 μl of 0.4 M Tris−HCl pH 8.4 containing 1 mM EDTA. (SDS, de-ionized urea or guanidine−HCl can be included to ensure solubility).
(ii) Add the solution to 10−20 mg of iodo-glass.
(iii) Incubate the sealed tube, under N_2, in the dark for 30 min at room temperature, with occasional agitation.
(iv) Add 20 μl of 2-mercaptoethanol and incubate for a further 15 min (to block excess alkylation sites).
(v) Wash the supports extensively with water and methanol.
(vi) Dry *in vacuo* and store at −20°C until required.

3.2.5 *Arginine*

Arginine residues represent a useful target for coupling since tryptic peptides can be produced which possess this residue at their C termini. Traditional methods utilize conversion of arginine to ornithine (16) which is achieved by incubation of peptide with 50% aqueous hydrazine (100 μl/100 nmol of peptide) at 70°C for 15 min. It is important to remove all traces of excess hydrazine by de-salting, dialysis or repeated freeze-drying. The ornithine-containing peptide can then be coupled to DITC-glass as described in *Table 8*. This approach suffers from two major problems. First, conversion to ornithine is only partially successful and secondly, hydrazine can also cleave sensitive peptide bonds. Less troublesome methods for arginine side-chain attachment are currently under development.

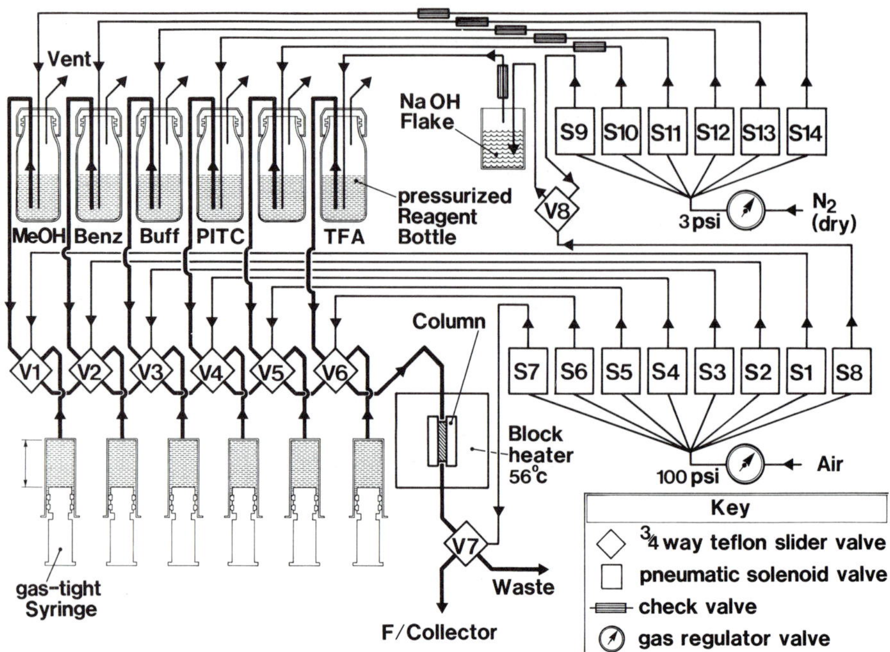

Figure 1. Diagrammatic representation of solid-phase sequencer. The column containing the glass-coupled peptide sits in an aluminium block heated to a constant 56°C. Reagents are supplied to the column from gas-tight syringes by way of 3/4 way Teflon slider valves, which are opened or closed by a set of pneumatic solenoid valves operated by a high-pressure (100 p.s.i.) air supply. The syringe pumps are electrically configured to refill from the reagent bottles when not delivering reagent to the column. A high-pressure solenoid valve also switches the column eluate between fraction collector and waste. A low-pressure (3 p.s.i.) supply of dry nitrogen, operating through a second set of solenoid valves, delivers nitrogen to the reagent bottles in rotation, maintaining them under a slight positive pressure of inert nitrogen. The nitrogen supply to the trifluoroacetic acid (TFA) bottle passes through a bottle of NaOH flake as an additional water trap. A slow venting of the reagent bottles ensures that temperature fluctuations do not result in over-pressurization of the bottles driving reagent back up the nitrogen lines. The timing of all valve and pump operations is controlled from an Acorn BBC Microcomputer.

4. INSTRUMENTATION

Figure 1 illustrates a diagrammatic representation of a solid-phase sequencer constructed in Leeds from standard commercially available components. It is included to demonstrate the relative simplicity of such machines, the basic features of which are present in all present and future solid-phase microsequencers. The operating limits of this equipment (1–2 pmol) are achieved by miniaturization of the reaction chamber and heating block (glass column of internal dimensions 0.8 × 25 mm equivalent to approximately 7 μl or 6–7 mg of glass beads), restriction of tubing volumes, rapid pulsing of reagents/solvents to eliminate inaccessible regions and introduction of a microprocessor controller. The reagents and solvents are kept under constant positive N_2 pressure.

The steps in a typical sequence cycle are given in *Figure 2a* and the solvents in *Figure 2b*. A typical cycle time is of the order of 22–24 min. Individual steps in the cycle can be varied depending on the particular N-terminal amino acid or combination of amino acids. This is often useful where particularly long sequences (up to 100 residues)

Figure 2. (**a**) Sequencing programme; (**b**) chemicals.

a

Time into cycle (min)	*Function*	*Duration* (sec)
0	Buffer	15
	Buffer + PITC[a]	20
	Wait (coupling)	85
2	Buffer + PITC	6
	Wait (coupling)	114
4,6,8,10,12,14	Repeat coupling step at 2 min intervals	
16	Methanol	60
17	Benzene	60
18	TFA	8
	TFA to fraction collector	5
	Wait (cleavage)	47
19	TFA to fraction collector	1
	Wait (cleavage)	59
20,21,22	Repeat cleavage step at 1 min intervals	
23	TFA to fraction collector	1
	Wait (cleavage)	49
23 min 50 sec	TFA to fraction collector	10
24	Methanol	60
25	Advance fraction collector	
	End cycle.	

b

Buffer:	50 ml Milli-Q water 10 ml 4-methylmorpholine (Fluka puriss p.a.) 150 μl *n*-propylamine (Fluka puriss p.a.) 140 ml methanol (Rathburn sequencer grade) 20% (v/v) aqueous TFA to give pH 8.8
PITC:	10% (v/v) *p*-phenyl isothiocyanate (Rathburn sequencer grade) in acetonitrile (Fluka puriss p.a.)
Methanol:	Rathburn HPLC grade
Benzene:	Fluka sequencer grade
TFA:	Anhydrous TFA (Rathburn sequencer grade) re-distilled from 1 mg/ml dithiothreitol; 5 μg/100 ml PTH-norleucine (Pierce) is added as an internal standard for HPLC.
Addresses:	Rathburn, Caberston Road, Walkerburn, Peebleshire EH43 6AD, Scotland, UK Fluka, Peakdale Road, Glossop, Derbyshire SK13 9XE,UK Pierce, 44 Upper Northgate St, Chester CH1 4EF,UK

[a]PITC, phenylisothiocyanate.

are required. The amounts of peptide/protein supplied to the machine can vary from 10 pmol to an upper capacity limit of over 3 nmol. *Figure 3* gives the results from three sequencing runs which illustrate the capacity, sensitivity and efficacy of this sequencing strategy. A new solid-phase microsequencer using glass and activated membrane supports has recently been launched by Milligen.

5. SEQUENCING PROTOCOLS

The peptide-coupled resins are packed into the reaction chamber of the sequencer which in our case is a small glass column. The packed resin is then subjected to a wash with methanol and the operating cycle started (see *Figure 2a*). Reagents and solvents should

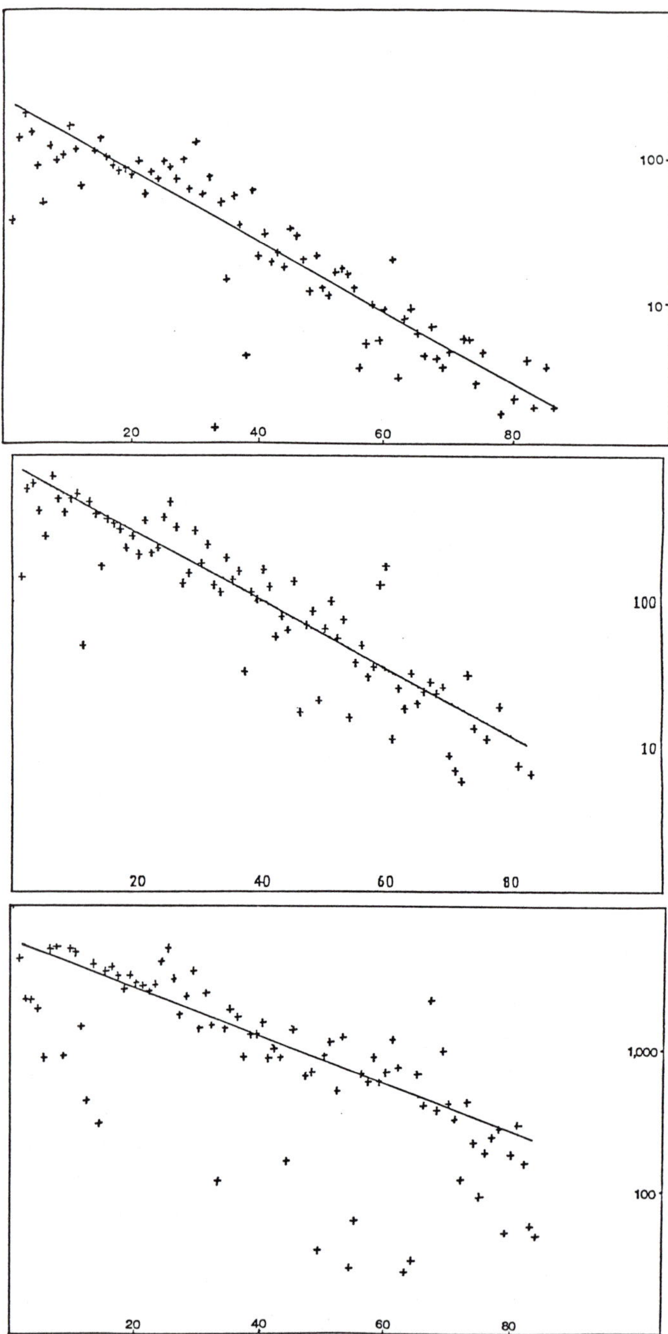

Figure 3. Examples of solid-phase sequencing runs illustrating operating range of equipment. **Top**: bacterioferritin (protein from Dr J.Smith and Professor P.Harrison, Sheffield University), 84 residues identified, initial yield 252 pmol, repetitive yield 94.6%. **Middle**: cyanogen bromide fragment of the A subunit of crustacyanin, 82 residues identified, initial yield 851 pmol, repetitive yield 94.8%. **Bottom**: intact A subunit of crustacyanin, 85 residues identified, initial yield 5.7 nmol, repetitive yield 96.3%. *x*-axis, residue number; *y*-axis, residue amount (pmol).

be of the highest purity available for the most efficient sequencing. However, one of the virtues of the solid-phase strategy is that lower quality reagents can be tolerated due to the efficiency of the washing procedures—with one exception. It is our experience that TFA is particularly critical to efficient microsequencing. For that reason we routinely re-distil this reagent (see *Figure 2b*) and store for only 1−2 months in sealed containers. In order to quench amino-reactive substances generated during the sequencing chemistry, *n*-propylamine is included in the buffer. This compound acts as a scavenger for amino-reactive groups and helps maintain the peptide amino terminus in its reactive unprotonated form. Some *N*-phenylthiocarbamyl (PTC)-propylamine is obtained along with the ATZ-amino acid but this is readily identified in the HPLC output and can be a useful indicator of some forms of machine malfunction. There are a number of properties of solid-phase sequencing which should be noted.

(i) The washing procedures which are used can be fairly rigorous and as a result samples are relatively free of chemical background noise. Modern equipment will yield 20−30 residues starting from 20 pmol of sequencing peptide.

(ii) Initial yields using carefully prepared supports are usually in the range of 40−70%. Repetitive yields range up to 97%.

(iii) The first residue recovered from material coupled using DITC is unreliable and, if important, its identity should be confirmed by another method. Depending on the coupling method used, lysine- (DITC coupling) or carboxyl-containing (carbodiimide and TFA anhydride coupling) residues are obtained in low yield. Positive identification of all these residues can be ensured by mixing supports which contain peptide coupled by two different methods.

(iv) Cysteine residues should be converted to stable derivatives (e.g. carboxymethylcysteine) prior to coupling since the unmodified amino acid is almost completely degraded.

(v) Since the cleaved ATZ-amino acids are washed from the reaction chamber with TFA, highly charged substituents, for example phosphorylated residues, can generally be recovered in reasonable yield.

(vi) Where contaminating peptides are present, it is often possible to irreversibly block these to further sequencing using *o*-phthalaldehyde (OPA). This reagent reacts irreversibly with primary amines but only to a small extent with proline (17). The consequence is that all peptides with the exception of those with an N-terminal proline can be rendered refractory to the Edman degradation procedure. The reagent can also be used to quench increasing backgrounds, even with pure peptides. A simple effective protocol is given in *Table 12*. When sequencing very large peptides or proteins the background will slowly rise again, probably due to low-level internal cleavages. The OPA procedure can then be repeated to extend the run.

(vii) The sequencing programme can be altered readily to effect efficient degradation. For example, proline residues can be subjected to extended cleavage times, hydrophobic stretches to repeated coupling steps, and acidic residues to a longer coupling step and a slightly reduced cleavage time.

(viii) The new generation of UV detectors are capable of handling material in the fmole range but there may be occasions where the chemical background masks the phenylthiohydantion (PTH)-amino acids. Under such conditions

Table 12. Reduction of PTH background using OPA.

The reduction of contaminating peptide background during sequencing may allow the sequence to be continued further than could otherwise be achieved. PTH background reduction can be achieved by blocking all N-terminal amino groups except proline by reacting them with *o*-phthalaldehyde (OPA). This can easily be done manually or by a semi-automatic programme.

1. Stop the sequencing run at the end of the appropriate cycle to leave proline as the new N-terminal amino acid.
2. Wash the sequencing resin with 5 ml of sequencing buffer (minus *n*-propylamine) containing 0.05% v/v mercaptoethanol and 5 mg of OPA at 56°C, in five aliquots over 20 min.
3. Wash the resin extensively with methanol, followed by benzene and then methanol again.
4. Re-start the automatic sequencing programme.

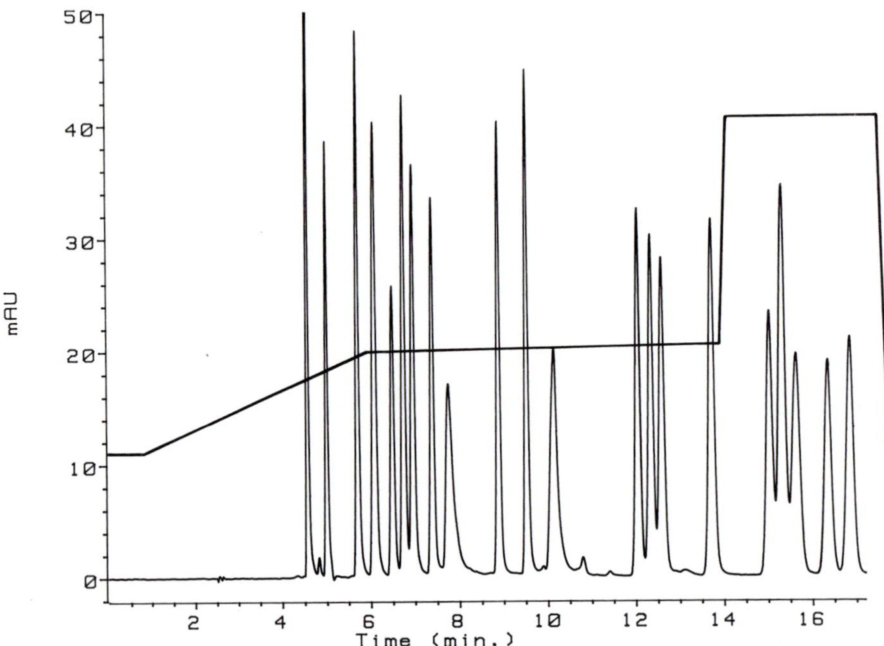

Figure 4. Separation of PTH-amino acids. Instrument:Hewlett Packard 1090M HPLC system; elution order: D,CM-C,E,N,S,Q,T,G,H,A,Y,R,M,V,P,W,F,K,I,L,Nle. Resin: Brownlee Spheri−5 RP-18 (220 × 2.1 mm); flow-rate: 0.21 ml/min; solutions: (**A**) Acetic acid, (2 ml/litre) pH 4.9 with NaOH—ionic strength and pH increased with column age to maintain elution position of arginine and histidine; (**B**) Acetonitrile (Far UV grade); loading solution; 30% (v/v) aq. acetonitrile, 1% (v/v) acetic acid. Detector: diode-array, collection of data at 269 and 313 nm, band width 30 nm.

Gradient:

Time (min)	*% A*	*% B*
0	90	10
0.01	78	22
1.0	78	22
6.0	60	40
14.0	60	40
14.1	20	80
17.0	20	80
17.45	90	10
17.5	STOP	STOP

6 min post-run equilibrium

4,*N*,*N*-dimethylaminoazobenzene-4′-isothiocyanate (DABITC) can be employed as the initial reactive isothiocyanate (see Chapter 6) substituted in place of the first two PITC pulses and followed by the remaining phenylisothiocyanate (PITC) pulses (see *Figure 2a*). Re-crystallized (from acetone), DABITC is used as a freshly prepared 0.5% solution in pure dimethyl formamide (DMF). The coloured DABTH residues can be detected by HPLC (18) or TLC (see Section 3.4, Chapter 6).

6. RESIDUE IDENTIFICATION

6.1 Conversion of ATZ- to PTH-derivatives

The ATZ-derivatives issuing from the column can be collected in a fraction collector or fed into an autoconverter for generation of the PTH-amino acids. Manual conversion can be achieved by first drying down the ATZ-derivatives *in vacuo*, followed by the addition of 50 µl of 30% (v/v) aqueous TFA and incubation under N_2 at 70°C for 15 min. The resultant PTH-amino acids are dried down under vacuum and re-dissolved in 30% aqueous acetonitrile/0.1 – 1% acetic acid in readiness for HPLC. Recoveries of all amino acids are near quantitative except for cysteine which is mostly destroyed, serine and threonine which are partly converted to their dehydro-derivatives and tryptophan which experiences limited destruction.

6.2 PTH identification

The HPLC profile at 269 nm is shown in *Figure 4* and the gradient profile is indicated in the legend. This separation has been achieved on both Du Pont and Hewlett Packard HPLC systems and with C18 resins from a number of manufacturers (Du Pont, Waters, Brownlee). For confirmation of serine and threonine residues, we routinely monitor the column eluate also at 313 nm, a wavelength at which their dehydro-derivatives absorb.

6.3 Quantitation

Quantitation is very important to sequencing and is ignored at one's peril. It is critical to establish that the initial sequencing yield roughly corresponds to the quantity of material coupled—a contaminant may well sequence while the main component is blocked, thereby giving misleading information. Again the yield of each residue should be carefully assessed during sequencing to ensure correct assignment, particularly in cases where the main residue is destroyed or for some reason not detected. Quantitation is easily achieved using modern integrating systems. Less common but very important is the ability of these programs to carry out sequential subtraction, that is the digitized subtraction of the preceding HPLC profile from the current one using pattern matching techniques. This has the very useful consequence of confirming the previous residue (negative peak) and of identifying the new residue (positive peak).

7. FUTURE PROSPECTS

It is quite clear that major developments are likely in the fields of sequencing chemistries, chemical coupling methods and insoluble supports. The blossoming of biotechnology alone has provided much of the impetus behind research in industrial as well as academic laboratories. It seems more likely than not that the solid-phase approach will emerge from this period of high research activity as a key feature in sequencing technologies whether they be liquid- or gas-phase systems. We can, therefore, look forward to ever-improving initial coupling and sequencing yields, ever higher sensitivities as fluorescence makes an impact and faster sequencing cycles as new chemistries are developed.

8. REFERENCES

1. Laursen,R.A. and Bonner,A.G. (1970) *Fed. Proc.*, **29**, 727.
2. Harris,E.L.V. and Angal,S. (eds) (1988) *Protein Purification Methods: A Practical Approach.* IRL Press, Oxford.
3. Findlay,J.B.C. and Evans,H.W. (eds) (1987) *Biological Membranes: A Practical Approach.* IRL Press, Oxford.
4. Laursen,R.A. (1971) *Eur. J. Biochem.*, **20**, 89.
5. Horn,M.J. and Laursen,R.A. (1973) *FEBS Lett*, **36**, 285.
6. Robinson,P.J., Dunnhill,P. and Lilley,M.D. (1971) *Biochim. Biophys. Acta*, **242**, 659.
7. Wachter,E., Machleidt,W., Hofner,H. and Otto,J. (1973) *FEBS Lett.*, **35**, 97.
8. Chang,J.Y., Creaser,E.H. and Hughes,G.J. (1977) *FEBS Lett.*, **78**, 147.
9. Weetall,H.H. (1970) *Biochim. Biophys. Acta.*, **212**, 1.
10. Aebersold,R.H., Teplow,D.B., Hood,L.E. and Kent,S.B.H. (1986) *J. Biol. Chem.*, **261**, 4229.
11. Laursen,R.A., Horn,M.J. and Bonner,A.G. (1972) *FEBS Lett.*, **21**, 67.
12. Previero,A., Derancourt,J., Colletti-Previero,M.-A. and Laursen,R.A. (1973) *FEBS Lett.*, **33**, 135.
13. Wittmann-Liebold,B. and Kimura,M. (1984) In *Modern Methods in Protein Chemistry.* Tschesche,H. (ed.), de Gruyter, Berlin New York, p. 229.
14. Davison,M.D. and Findlay,J.B.C. (1986) *Biochem. J.*, **234**, 413.
15. Davison,M.D. and Findlay,J.B.C. (1986) *Biochem. J.*, **236**, 389.
16. Morris,H.R., Dickinson,R.J. and Williams,D.H. (1973) *Biochem. Biophys. Res. Commun.*, **51**, 247.
17. Brauer,A.W., Oman,C.L. and Margolies,M.N. (1984) *Anal. Biochem.*, **137**, 134.
18. Lehmann,A. and Wittmann-Liebold,B. (1984) *FEBS Lett.*, **176**, 360.

CHAPTER 4

Gas- or pulsed liquid-phase sequence analysis

M.J.GEISOW and A.AITKEN

1. INTRODUCTION

Gas–liquid or pulsed liquid-phase sequencers represent technical developments of the standard Edman–Begg automated equipment (1). Instead of the sample being maintained in the liquid-phase in a spinning cup, it is physically entrained, with or without a polycationic carrier, in a glass microfibre filter (2). In the former case ionic and hydrogen bonding interactions also help to maintain the polypeptide chain *in situ*. Washout of protein by polar reactants is prevented by supplying these either in the vapour-phase or as metered pulses of liquid, which are subsequently removed from the filter by evaporation.

Further improvements in performance are achieved by miniaturization of the reaction chambers and valves. The latter are based on an original design of Wittmann-Liebold (3). Emphasis has also been placed upon purification of reagents and solvents. The increased yield of these sequencers has led to the development of microbore high-performance liquid chromatography (HPLC) systems for identification and quantitation of the phenylthiohydantoin (PTH)-amino acids. With such equipment, routine sequencing of 100 pmol amounts is possible, with detection limits down to 1 pmol or less.

The possibility of convenient automated sequencing of such low amounts imposes unusual demands upon sample preparation and handling. Sample transfers alone can result in major losses at the picomole level. Accordingly, much emphasis is placed upon sample work-up, both in this chapter and others dealing with microsequencing.

2. SEQUENCER MECHANICS AND CHEMISTRY

Commercial equipment based upon the gas–liquid solid-phase sequencer of Hewick *et al.* (2) is now available and several laboratories have constructed their own machines. Methods and comments in this chapter specifically apply to machines manufactured by Applied Biosystems Inc. because these are in regular use in the authors' laboratories. The majority of the remarks will nevertheless apply generally to machines employing the same principles.

A multi-purpose protein sequencer incorporating a programmer, inert gas system, valves and conversion flask has recently been developed by Reimann and Wittmann-Liebold (4). The reaction chamber is interconvertible to enable solid-, liquid- or vapour-phase technologies in the Edman chemistry. This system is now commercially available (Knauer Gmbh, Hegauer Weg 38, D-1000 Berlin 37, FRG).

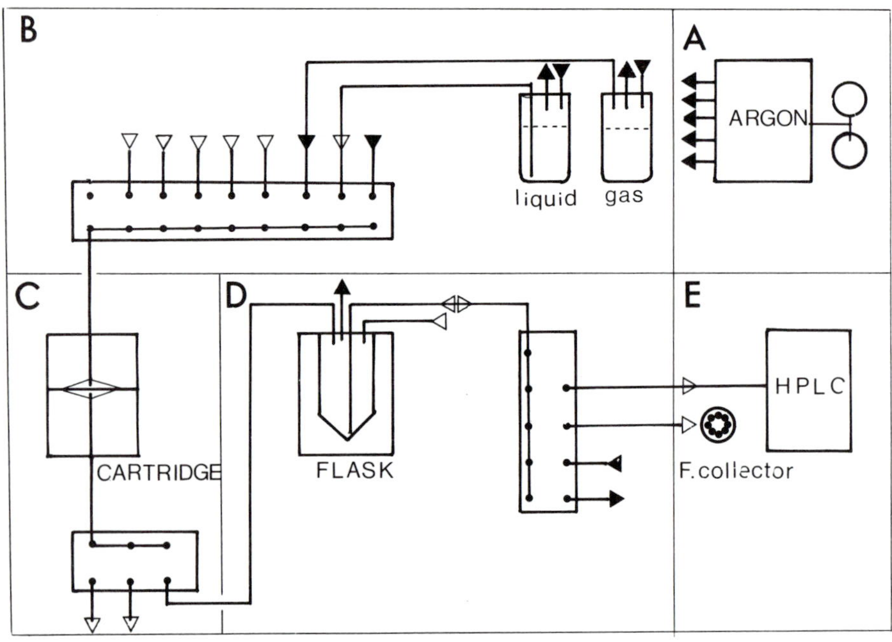

Figure 1. An outline of gas (►) and fluid (▷) transfers in gas–liquid-phase sequencers.

2.1 Design of gas–liquid-phase sequencer

The mechanics of these sequencers can be divided into a small number of systems as indicated in *Figure 1*. Reagent and solvent delivery and removal, ATZ-amino acid and PTH-amino acid transfers are made by argon supplied from a buffered distribution system (A). Zero grade argon (99.998%) is scavenged of moisture and particles (>0.4 μm) by traps and supplied at a backing pressure of 40–60 p.s.i. Argon is used in preference to nitrogen since it is a denser gas (forming a better inert 'blanket') and can be obtained in a higher state of purity. In the gas-phase instrument, regulators control the pressure in the head space of each reagent or solvent reservoir to a dynamically pre-set value. The argon is vented through a relative constriction from the head space during adjustment and subsequent timed delivery. In the pulsed-liquid sequencer, the metering of reagent is made quasi-independent of pressure, by overfilling a loop of defined volume prior to switching the loops content to the reaction chamber. Reagent deliveries are made via a common path which is swept by argon. The reagent and solvent delivery system (B) contains reservoirs which all deliver liquid except for the gas-phase instrument where trimethylamine (TMA): water (R2) and trifluoroacetic acid (TFA, R3) vapours are led from the head space to the sample cartridge. The effluent argon streams are vented into the headspace of a waste vessel, to which is also routed the argon/reagent or solvent vapour vented from the headspaces of each reservoir. This enters an exhaust manifold which is swept by nitrogen and the gaseous waste is partly scrubbed of trimethylamine vapour over a concentrated acid (50% H_3PO_4) and routed to fume extract.

The thermostatically-controlled cartridge (C) consists of hard glass cylinders drilled with an axial capillary connecting a double conical cavity to inlet and outlet lines. Support

Table 1. Gas–liquid sequencer reagents.

Symbols	*Chemical*	*Normal delivery*	*470A gas pressures*	*477A liquid delivery*
R1	5% Phenylisothiocyanate in *n*-heptane	30 μl	0.6 p.s.i.	
R2	12.5% Trimethylamine in water	5 ml/min	0.5 p.s.i.	
R3	Anyhdrous trifluoroacetic acid; 0.001% DTT	5 ml/min	0.5 p.s.i.	22 μl
R4	25% Trifluoroacetic acid in water; 0.001% DTT	50 μl	1.5 p.s.i.	
S1	*n*-Heptane (optional 0.001% DTT)	0.8 ml	1.35 p.s.i.	
S2	Ethyl acetate (optional 0.001% DTT)	1.6 ml	1.2 p.s.i.	
S3	*n*-chlorobutane (optional 0.001% DTT)	1.2 ml	1.5 p.s.i.	
S4	20% Acetonitrile in water; 0.001% DTT	120 μl	1.5 p.s.i.	

for a Whatman GF/C type filter (1.2 cm) in a machined recess is provided by a porous Teflon filter which also seals the cartridge blocks against leaks or oxygen entry. The outlet of the sample cartridge is routed either to waste, to a valve permitting direct collection of unconverted ATZ-amino acids or to the conversion flask. Apart form the latter connection, cartridge and flask chemistries can proceed independently and asynchronously under programme control.

The conversion flask (D) is independently thermostatically controlled (~55°C in the gas-phase or 64°C in the pulsed liquid instrument). A Teflon line is used to pick up the flask contents from its conical bottom or to bubble argon to mix the contents. The remaining lines pressurize the headspace for PTH transfers or washes and deliver the ATZ-amino acid or conversion chemicals or solvents. PTH-amino acid standards delivered from a reservoir to the flask can be exposed to the same physical and chemical pathways as ATZ-amino acids and thus act as better qualitative and quantitative standards for subsequent HPLC analysis.

The PTH-amino acid solution in the flask is either delivered directly to a fraction collector (E) or dried and reconstituted with a constant volume of acetonitrile/water (20%). This is metered by a fixed volume loop. Argon drives this PTH-amino acid solution into the sample loop of a synchronized HPLC system. Loading is controlled by timing and a constriction on the effluent side of the sample loop. This constriction slows down subsequent liquid flow when the loop has filled.

2.2 Sequencer chemistry

Table 1 lists reagents used in the gas (470A) and pulsed liquid (477A) sequencers. Not all commercially-available reagents of 'sequencer grade' quality may allow sequence determination to the current limits of sensitivity. The most crucial reagents are phenylisothiocyanate (PITC) and TMA. Problems may also be encountered with some batches of extraction solvent, ethyl acetate. The most stringent test is sequence analysis of picomole amounts of a standard protein (e.g. myoglobin or β-lactoglobulin).

2.2.1 *Purification and storage of trimethylamine and polybrene*

(i) *Trimethylamine (TMA).* TMA is refluxed with phthallic anhydride at 5°C for 2 h. The resulting solution is distilled through phthallic anhydride into a receiving flask in dry ice–acetone. TMA is mixed with HPLC grade water to give a 12.5% v/v solution and stored under an argon atmosphere at 4°C. At room temperature, TMA gradually decomposes to generate dimethylamine (DMA). In the reaction cartridge DMA reacts with phenylisothiocyanate (PITC) to produce *N*-dimethyl-*N'*-phenylthiourea (DMPTU). DMA is more volatile than TMA and will be preferentially removed in the stream of argon. It may build up, however if the sequencer is infrequently used.

(ii) *Polybrene (1,5-dimethyl-1,5-diazaundecamethylene polymethobromide).* The average molecular weight of this polymer should be 20 000 daltons. The main contaminants are short chain and unpolymerized material. In the most sensitive sequencing runs, polybrene-treated filters are pre-cycled before loading sample and lengthy clean-up of the polybrene may not be worthwhile. A special pre-cycling programme is either supplied, or can be easily devised, for any instrument. Dissolving the commercial product at 200 mg/ml, clarification by centrifugation at 20 000 *g* for 30 min, followed by extensive dialysis against HPLC grade water has proved effective. Care must be taken to allow for expansion of the dialysis tubing.

Dialysis against activated charcoal suspension has proved effective in removing significant unknown UV-absorbing materials. The dialysate is used directly or lyophilized and reconstituted in HPLC grade water containing AristaR grade sodium chloride at 7 mg/ml at a final polybrene concentration of 50 mg/ml. The salt is claimed to improve extraction of ATZ-Asp, Glu, His and Arg by *n*-chlorobutane.

Glass-fibre discs can be prepared in advance with polybrene *in situ*. Sheets of GF/C are washed with TFA as described in Section 3 (sample application). Pre-purified polybrene is applied to the sheets (50 mg/ml) in water, ref. 5). The sheet is suspended and air-dried briefly before a final distilled water wash and drying *in vacuo*. Discs are cut from the dried sheets by a 12 mm cork borer.

2.3 **Sequencer optimization**

The surest diagnostic of performance is to run a known standard at about 100 pmol for around 10 amino acids. Records of initial and step repetitive yields should be kept for future reference. A long synthetic peptide can be a good test sample.

2.3.1 *Optimization of reagent and solvent deliveries*

(i) The delivery of R1 (PITC) should be just sufficient to wet the filter. No PITC solution should appear in the cavity below the filter disc. A delivery of 30 μl of 5% (v/v) PITC in heptane gives about 1.5 mg (10 000 nmol). Even this amount of PITC is a considerable excess. The minimum should be a 500-fold molar excess over free α-amino group. The amount delivered can be determined by extracting the filter with *n*-heptane and measuring the absorbance at 254 nm. If multiple filters are placed in the cartridge, the R1 delivery must be increased in proportion to the number of filters.

(ii) Delivery of R2 (TMA) should not lead to visible moisture appearing below the filter, which would indicate potential loss of sample. On the gas-phase instrument a headspace pressure of 0.5 p.s.i. during delivery of R2 is optimal. Exhaustion of the TMA will be close when the level of liquid in the reservoir has fallen by only 10%. The initial volume should therefore be marked on the vessel together with the calculated position representing 90% of this initial volume.

(iii) Delivery of S1 (*n*-heptane). Excess (unreacted) PITC is dissolved out of the filter to waste. Continued exposure of PITC to water, both in the cartridge and especially in the conversion flask leads to high levels of *N,N'*-diphenylthiourea (DPTU). The latter is formed after PITC reacts with water to produce hydrogen sulphide, carbon dioxide and aniline. The latter reacts with PITC to form DPTU.

(iv) Delivery of S2 (ethyl acetate). Excess DPTU and other by-products are dissolved out of the filter to waste. Excessive washing with S2 is likely to lead to loss of small peptides, especially hydrophobic ones.

The deliveries of both S1 and S2 are best adjusted in relation to the level of artefact peaks appearing in the HPLC traces. The main by-products of the Edman chemistry are DPTU, DMPTU and *N,N'*-diphenylurea (DPU). The last by-product forms by oxidative desulphurization of DPTU and is normally present at levels of 5–10% of DPTU. Provided that each of these by-products are well resolved from PTH-amino acids in the HPLC system used, it may be better to reduce S2 delivery, minimizing the opportunity for peptide loss. However, unusual levels of any of these by-products with normal S2 delivery may indicate trouble, as outlined below.

(a) *High levels of DPTU.*

(1) Lead in R1 delivery valve (excess PITC delivery to flask).
(2) Water contamination in R1 (e.g. an improperly dried reservoir).
(3) Expired or faulty argon moisture trap (check silica gel indicator).
(4) Poor quality argon (the above applies).
(5) Polybrene is highly hygroscopic and retains water from R2. The level of DPTU will increase in proportion to polybrene loading. Very low DPTU levels result from loading proteins without polybrene or from electroblotted proteins (see Chapter 1).

(b) *High levels of DMPTU.*

(1) Poor quality or improperly stored TMA.
(2) Old or infrequently-used TMA on the sequencer (thorough venting with argon may reduce DMPTU levels, since DMA is more volatile than TMA).

(c) *High levels of DPU.*

(1) Contamination of R1 with air (phenylisocyanate formed from PITC).
(2) Major air leak (in delivery lines to cartridge or poor cartridge seals (including the Teflon filter support seal).
(3) Poor quality argon.

(v) Delivery of S3 (*n*-chlorobutane). The initial delivery of S3 should be sufficient to fill the cavity and wet the filter, but not appear below the filter (seen as a phase-dense meniscus against the rear illumination). This will allow precipitation/penetration of potentially extractable hydrophobic samples. A more critical factor is the presence of residual TFA from the previous cleavage step. If the S3 passed straight through before allowing time for residual TFA to become diluted, loss of peptide material may occur. The subsequent transfer with S3 should be relatively slow (should not appear in the conversion flask before 20 sec following initiation of the transfer function). Delivery is made directly into 25% TFA in water so that conversion starts immediately. This procedure also avoids exposing the unstable ATZ-derivatives directly to the hot glass surface and introduces dithiothreitol (DTT) into the flask. A method of conversion using dry conditions and higher temperature has been attempted and better recoveries of some derivatives such as those of Ser and Thr can be achieved. This method is not the one of choice however, since poorer recoveries of other PTH-amino acids result (e.g. PTH-His). Since the charged ATZ-amino acids are hardest to extract, optimization of transfer could be effected using a peptide rich in one of these (i.e. Asp, Glu, His or Arg).

2.3.2 *Metering of solvents and reagents*

Sufficient time must be allowed for volumetric loops to slightly overfill, to allow reproducible metering. Because the ends of the loops are hidden in compression fittings, this overfilling can be difficult to assess. However, by monitoring the waste lines during timed loop filling, it should be possible to arrive at an adequate time determination.

2.3.3 *Drying times*

Drying functions on the sequencer are critical at certain stages. In general, drying should be minimized, since unstable ATZ- or PTH-derivatives are brought into chemical contact with hot silica surfaces leading to losses. In the filter, the extractability of ATZ-His by *n*-chlorobutane may be adversely affected by overdrying. With the on-line HPLC operation of the ABI sequencer it is critical that the PTH-amino acid in the conversion flask is completely dry before the flask, sample line, sample loop and effluent restrictor are cleared by argon gas. Drying times may drift substantially, especially for the conversion flask. In some cases this appears to be a combination of creep of the Teflon lines, altering their resistance to gas flow and build up of insoluble deposits in the lines. Some of this material may be contaminants from the reservoirs. Talc from protective gloves may be one source, another may be re-precipitated silica from HPLC columns. The only remedy is replacement of the affected Teflon lines or re-cutting the end of the conversion flask tubing.

2.3.4 *Transfer time to on-line HPLC*

The timing for argon-driven transfer of the re-dissolved PTH-amino acid from the conversion flask to the HPLC sample loop is critical. A tolerance on this transfer is obtained by connecting a 0.007 inch restrictor tube to the Rheodyne effluent port. The transfer time is set 2 sec less than the time of appearance of sample at the end of this restrictor. The problems appear to be partial blockage of the restrictor, crimping of the Teflon connection between the sequencer and HPLC or leakage at the compression fitting in the valve block. Insufficient drying of the PTH-amino acid can leave fluid blocking the restrictor. The restrictor is also susceptible to blockage by small particles. It can be cleared by connecting it to the outlet of an HPLC pump and forcing filtered solvent through it at high flow-rate.

Excessive amounts of detergent loaded with the sample may cause foaming in the transfer lines. The resulting menisci will slow delivery of the PTH-amino acids due to increased surface tension/pressure in the Teflon lines.

3. SAMPLE APPLICATION, PRE-TREATMENT AND SPECIAL PROGRAMMES

The support medium is Whatman GF/C glass fibre or polyvinylidene difluoride (PVDF). Proteins can be transferred to this support, with or without manipulation of the surface charge by Western blotting polyacrylamide gels (see Chapter 1). Before use in the sequencer, filter discs should be cleaned with sequencer-grade TFA.

(i) Place up to 50 filter discs in 8 ml of TFA in a glass scintillation vial (cover the foil lid liner with a Teflon disc such as those in solvent bottle tops).
(ii) Swirl the filters to remove bubbles and leave at room temperature for 2 h.
(iii) In a fume hood, aspirate the TFA and place the filters on clean glass Petri dishes with Teflon-tipped tweezers.
(iv) After air drying, remove residual TFA in a vacuum desiccator containing potassium hydroxide pellets (leave overnight).
(v) Store the filters in a clean capped scintillation vial or polyethylene vial insert.

Normally, 30 μl (1.5 mg) of polybrene is pipetted from stock stored aseptically at 4°C. For sequencer loadings of less than 500 pmol, the polybrene should be pre-cycled before sample application. Larger applications may be applied directly to the polybrene-treated filter after drying. Alternatively the discs may be pre-treated with polybrene as described above. The pre-cycling should commence with a coupling step (not cleavage) since the polybrene could otherwise be dislodged, precipitate with butyl chloride and lead to blockage of transfer lines. Some of the sample that is adhering to this polybrene could also be removed. Sample and polybrene may be loaded together and sequencing commenced immediately, but for amounts less than 100 pmol, filter conditioning by means of a pre-cycle programme is advisable.

After complete sequencing of a relatively short peptide on a pre-cycled polybrene-treated disc, the background level of PTH-amino acids should fall to virtually zero. In this case a new sample can be loaded directly onto the disc and sequencing re-commenced immediately. In the author's laboratory this process has been repeated for up to four peptides in succession. It may in fact be a positive advantage to follow sequencing of a short peptide in this way with a protein or large fragment that is available in very low amounts. All impurities on the disc that would react with the amino groups on the second sample will have already been 'mopped up'. A better initial yield at high sensitivities has been found. This is also very economical in time and money.

The sample should ideally be in 40–100 μl salt-free homogeneous solution and the final purification steps should aim to achieve this (see Chapter 1). Essentially, the final step should introduce volatile solvents and buffer salts and minimize drying, dialysis or concentration steps. Avoid, if possible, complete drying; reduction to a small volume often prevents solubility problems.

If the protein or peptide is supplied dried from the last stage of purification, it should be re-dissolved in 30–60 μl of 20% aqueous acetonitrile. Peptides purified by HPLC are often dried in a vacuum centrifuge from ammonium bicarbonate or TFA-buffered solvent in polyethylene Eppendorf tubes. It is worth re-dissolving this peptide in a buffer whose non-polar solvent composition approximates to that of the gradient in which it eluted. Good solvents for dried protein or peptide are 1 M HCl, 1–100% TFA, 50% glacial acetic acid or 70% formic acid. Protein can be loaded in non-ionic detergent (e.g. 1% Triton X-100). Ionic detergents (sodium dodecyl sulphate, SDS) create problems at high loadings, For 1.5 mg polybrene, no more than 20 μl of a 0.05% (w/v) SDS solution should be applied. More SDS might be loaded if the polybrene is increased in proportion. If the sample is loaded in detergent (ideally no more than 1–2%) the disc should be resoaked in 30 μl of 25% TFA after the original sample has dried. Solutions of urea, guanidinium-HCl, primary or secondary amines should be avoided. If more than 1% SDS is present, this can lead to foaming during transfers. This would for example lead to the appearance of large numbers of menisci in the delivery line from conversion flask to on-line HPLC and radically alter the delivery time. SDS should always be re-crystallized from ethanol:H_2O (6) (Serva or Bio-Rad are good sources of starting material).

Partial filters can be arranged upon the Teflon support filters. If direct blotting onto glass fibre or PVDF leads to lower than expected recoveries, it may help to place a pre-cycled polybrene-treated filter directly below. Polyacrylamide gel electrophoresis may be carried out with a reversible cross-linker such as bis-acrylylcystamine (7). The excised gel band is treated with high levels of 2-mercaptoethanol or DTT and the liquified extract passed through clean-up procedure such as an Aquapore RP-300 reversed phase column.

3.1 **Sequence analysis of resin-bound synthetic peptides**

A sample of such peptides may be analysed by removing a few beads of resin from a peptide synthesizer and placing on a pre-wetted disc (Solvent S4). For optimal results a special programme should be employed that has extended cleavage and transfer times. Excellent results have been achieved in the author's laboratory. This technique is very successful in monitoring the progress of long syntheses.

3.2 **Solid-phase coupling of peptides**

This has been carried out in the author's laboratory by coupling peptides to activated glass fibre supports—aminopropyl glass with diisothiocyanate (DITC) coupling—carried out in a similar manner to that described in Chapter 3 (work done in collaboration with Dr A.D.Auffret, Pharmacia LKB Biochrom).

3.3 **Sample de-salting and clean-up**

A protein may be in an inconveniently large volume of salt. If so, it may be possible to recover concentrated, salt-free material using the methods outlined in Chapter 1. For example, ion-exchange chromatography using ammonium acetate; a pH step on an HPLC ion-exchange cartridge or reverse-phase step elution with propanol at pH 7. Electro-concentration in an electroelution cell might also be considered. It these methods do not apply then protein can be lyophilized at the bottom of a cooled vacuum centrifuge attached to a good cold trap and vacuum source. Alternatively, protein can be precipitated by addition of ethanol to 90% (v/v) for 4–6 h at −20°C; collected by centrifugation (14 000 *g*, 15 min, 20°C) and washed with 95% ethanol at 4°C. Precipitation with chloroform is not recommended since impurities that react with amino groups may be present. All residual solvent is aspirated with a drawn out glass pipette. A magnifying glass will be useful in this and subsequent manipulations.

3.4 Reduction and S-alkylation of proteins

There are a large number of cysteine alkylating agents available but 4-vinylpyridine offers several advantages for the gas-phase sequencer (8).

(i) Dissolve the protein in 50 μl 6 M guanidinium−HCl, 0.5 M Tris−HCl, pH 7.9, 1 mM ethylenediamine tetraacetic acid (EDTA) by vortexing and incubating at 37°C until no obvious pellet reforms on centrifugation.
(ii) Add high purity DTT (Calbiochem) to 10 mM final concentration.
(iii) After 1 h at 37°C, add 4-vinylpyridine dissolved in acetonitrile to a final concentration of 20 mM (2-fold molar excess over DTT).
(iv) After 2 h at room temperature, free alkylated protein of guanidinium−HCl by re-precipitation with 90% (v/v) ethanol as above, and again recover the protein by centrifugation.

Reduced, alkylated proteins probably will not behave well on the chromatographic steps suggested above so that there are fewer alternatives to the above precipitation step. De-salting of the protein may be carried out by gel exclusion or wide-pore reversed phase HPLC. Use of radiolabelled alkylating agent may be useful in this respect. Many of the above problems can be circumvented by treatment of the protein on the glass fibre disc with 4-vinylpyridine (in the presence of tributylphosphine) in the vapour phase (9). The pyridylethylated protein or peptide is then sequenced directly, after a special solvent wash cycle.

3.5 OPA blocking

When background levels of amino acids build up during a sequencer run, these may be radically reduced by interrupting the sequencer run and introducing a reagent specific for primary amino groups. The reagent of choice would appear to be *o*-phthalaldehyde (OPA). If this is introduced (while proline is at the amino terminus of the polypeptide chain of interest) this will remove all the sequenceable amino acids except the prolylamino-terminal sequence (10). The OPA treatment should be followed by a PITC coupling step otherwise preview may be seen—presumably due to formation of an OPA−prolyl adduct that may be cleaved by anhydrous acid if the blocking is immediately followed by a cleavage step. The new Applied Biosystems model 477A protein sequencer has an extra reagent bottle for cartridge functions (as well as one for conversion functions). This makes development of a blocking programme more simple. We have otherwise carried out OPA blocking on the model 470A sequencer using bottle S2 containing OPA (0.5 mg/ml in butyl chloride). S1 contains a mixture of heptane:ethyl acetate, 1:1.

3.6 Special programmes for optimization of particular amino acids

These may be used for the following amino acids if a preliminary sequencing run has suggested their presence at particular cycles.

3.6.1 *Proline and glycine*

An extended reagent R3 delivery (TFA) and elevated temperature in the reaction cartridge may be employed to optimize cleavage of the amino acids during the cycles at which they are expected to appear. Use of these conditions at all cycles may lead to higher background due to polypeptide chain cleavage.

3.6.2 *Serine*

Cartridge temperature may be reduced and pre-coupling base (R2) delivery increased. This may help to minimize amino-terminal blocking and polypeptide chain cleavage that can occur at seryl residues.

3.6.3 *Glutamine*

Cartridge temperature and cleavage conditions could be reduced at the cycle before glutamine is expected, to help prevent cyclization of this amino acid to pyrollidone carboxylic acid.

4. PTH-AMINO ACID ANALYSIS

HPLC detection and quantitation of amino acids is almost universal in microsequencing. Isocratic systems have been described by Lottspeich (11) and Ashman and Wittmann-Liebold (12). Many more gradient systems have been reported. The gradient system reported in *Table 2* (13) has been used in the authors' laboratories either with reciprocating piston pumps or with syringe pumps which form part of the Brownlee laboratory MPLC equipment.

4.1 **Separation**

The separation of PTH-amino acids, DPTU and DMPTU is shown in *Figure 2. Table 3* lists the relative positions of other common amino acid derivatives and artefacts. Most of these are not specific to gas- or pulsed liquid-phase sequencers. Beta-elimination of the hydroxyl groups of Ser and Thr leads to formation of the dehydro-derivatives in the conversion flask. These react with DTT present in R4 (*Table 1*) to give relatively stable adducts. Threonine gives rise to at least four characteristic peaks in addition to the unmodified PTH-derivative.

Table 2. Reverse-phase HPLC of PTH-amino acids.

Column	Brownlee PTH C18 (22 cm × 2.1 mm or 4.6 mm)
Solvent A	5% THF in water 100 mM ssodium acetate pH4 (±0.3)
Solvent B	Acetonitrile
Flow-rate	1.0 ml/min (4.6 mm column) Temperature 55°C ± 3°C 0.2 ml/min (2.1 mm column)
Detector	270 nm (variable) 250 nm (fixed wavelength)
Gradients	

4.6 mm column		2.1 mm column	
%B	Time (min)	%B	Time (min)
8	0	10	0
19	2	14	2
51	16	40	20
51	19	60	25
80	19.1	10	25.1

5. PHOSPHOAMINO ACIDS

The identification of phosphoamino acids is a very important aspect of primary structural studies of proteins. Very poor recovery of phosphoserine, -threonine and -tyrosine derivatives are achieved in gas-phase (or pulsed-liquid) sequencers and spinning cup sequencers (14,15). The best recovery of phosphoamino acids is achieved with solid-phase sequencers (16). One method has been developed (17) which involves β-elimination and addition of methylamine. β-elimination followed by reduction with sodium borohydride to convert phosphoserine to alanine (18) has also been used. *O*-glycosylated serine residues can be similarly converted to alanine (19). None of these methods has proved particularly useful in practice especially when multiple phosphorylation sites are encountered (as is very common). A method has recently been described (20) that employs cutting the disc into pieces (after loading the radiolabelled phosphopeptide) and removing one piece after each cycle in which phosphorylated residues may be expected. The residual peptide is extracted with 50% formic acid and subjected to reverse-phase HPLC. The radioactivity will co-elute with the peptide until the phosphorylated amino acid(s) is/are reached when the ^{32}P radioactivity (which has been converted to inorganic phosphate) will elute in the void volume.

5.1 **Phosphoserine-containing peptides**

The following method and its variants are specific for phosphoseryl residues. Modification of phosphoserine in peptides by β-elimination of the phosphate followed by addition of ethanethiol leads to the stable adduct S-ethylcysteine (21,22).

(i) Dissolve the peptide in a capped tube containing 0.2 ml of water, 0.2 ml of dimethylsulphoxide, 80 μl of ethanol, 65 μl of 5 M sodium hydroxide and 60 μl of 10 M ethanethiol.

(ii) Flush the tube with nitrogen (easily achieved by needles inserted through the plastic lid) and incubate for 1 h at 50°C.

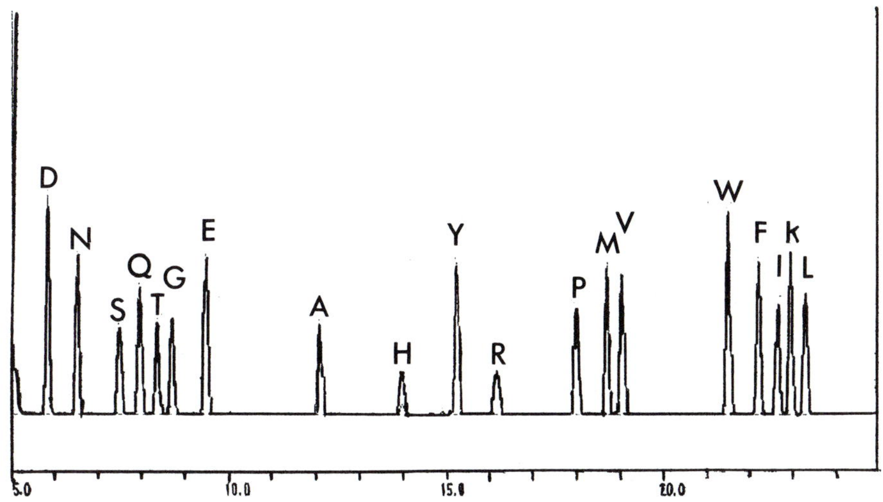

Figure 2. Elution of PTH-amino acids (Pierce Chemical Co.) on an Applied Biosystems 120A PTH analyser using the chromatography conditions described in the text.

Table 3. Elution of PTH-amino acids and artefacts from Brownlee PTH C18 reverse-phase column.

Figure 2 together with this table indicate the relative elution position of PTH-derivatives and side products using a Brownlee PTH C18 reverse-phase column (220 × 2.1 mm, 5 μm particles) and gradients of aqueous tetrahydrofuran (5%, buffered to ~pH 4 with sodium acetate and acetonitrile).

Amino acid derivative or reaction product
[a]Me-Asp and Me-Glu
Cysteic acid
DTT (reduced)
DTT (oxidized)
Asparaginyl *N*-acetylgalactosamine
Aspartic acid
Asparagine
Serine
S-carboxylmethylcysteine
Glutamine
Threonine
Glycine
Glutamic acid
DMPTU
Histidine
Succinyllysine
Hydroxyproline I
Alanine
Hydroxyproline II
Histidine
[b]DTT adduct of dehydroalanine (Serine)
Arginine
Tyrosine
Arginine
[b]DTT adduct of dehydro-α-aminoisobutyric acid (threonine) 2−4 peaks
Proline
Methionine
Valine
4-pyridethylcysteine (PE-Cys)
Hydroxylysine
S-ethanolcysteine
DPTU
DPU
Tryptophan
Phenylalanine
DPU
Isoleucine
Lysine
Leucine
Methyllysine
Trimethyllysine
S-ethyl cysteine
[c]Norleucine

(0.5 ml of 12.5% aqueous TMA per litre of buffer A (1 mM), will sharpen and bring forward the elution of the PTH derivatives of His, Arg and PE-Cys.

[a]Methanolic HCl conversion appears to be virtually obsolete on instruments with on-line detection.
[b]The DTT adducts of serine and threonine may be monitored at an additional wavelength, 313 nm.
[c]On-line detection methods of PTH analysis do not normally employ internal standards such as PTH-norleucine. The levels of DMPTU and DPTU during a sequencing run could instead be used to ensure that the delivery of products is consistent. Sudden alterations in their levels should be viewed with suspicion.

(iii) After cooling, add 10 μl of glacial acetic acid.
(iv) Apply the derivatized peptide directly to the sequencer or concentrate by vacuum centrifugation.

The S-ethylcysteine elutes just before DPTU. Some DTT adduct of PTH-Ser is also seen, resulting from β-elimination during conversion of ATZ-S-ethylcysteine. Independent confirmation of the presence of phosphate in a given peptide should be sought, because *O*-glycosylated serine would also undergo β-elimination during the derivatization procedure. In this case, however, sequence analysis of the unmodified peptide would not give a PTH-amino acid at the position of the glycosyl residue, whereas a substantial amount of the DTT adduct of serine would be expected in the case of a phosphorylated residue.

6. SUPPLIERS

6.1 Sequencer reagents

Rathburn, Peebleshire, Scotland, UK
Burdick and Jackson, USA
Applied Biosystems—complete sequencing reagent kits for 1000 to 1500 Edman cycles as well as individual chemicals

6.2 Associated equipment

Argon 99.998%, British Oxygen Company Special Gases or Air Products; Univap or Gyrovap concentrator, Uniscience or V.A.Howe Ltd; Spectrol regulators (Type 51B GG BS3) for zero grade argon, British Oxygen.

6.3 General chemicals

TFA for pre-treating glass fibre, Pierce-Warriner; Polybrene for electroblotting, Aldrich; iodoacetic acid, British Drug Houses; 4-vinylpyridine, Aldrich; N-ethylmorpholine, Aldrich; Gold Label dithiothreitol, Calbiochem; high purity grade trimethylamine, Eastman Organic Chemicals, Rochester, NY.

6.4 HPLC solvents

Acetonitrile, Rathburn; Romil; Fisons (last source may be the best for high sensitivity sequencing needing low absorbance at 254 nm)
Tetrahydrofuran, Aldrich (no stabilizer—UV-active!)
Sodium acetate, BDH HPLC grade

6.5 Sequencing support media

Glass fibre sheets, Whatman (either GF/F or GF/C)
Polyvinylidene difluoride (PVDF) 'Immobilon', Waters-Milligen
Porous Teflon disks, Zitex filter membrane extra coarse, Chemplast Inc. Wayne, NJ, USA

6.6 Fraction collector tubes

Durham tubes (6 × 30 mm) soda glass tubes can be used in fraction collector of model

470A Applied Biosystems only (not 477A) instead of the much more expensive Waters insert vials. When using on-line HPLC detection it is extremely rare to re-inject samples. There must however be a tube at the appropriate position for the sequencer to continue. The remaining sample is just as easily recovered for measuring radioactivity, etc.

6.7 **Instruments**

Pulsed liquid-phase protein sequencer models 477A (with gradient on-line 120A HPLC) and 471A (with isocratic on-line HPLC), (gas-phase model 470A is no longer supplied), Applied Biosystems Ltd, Kelvin Close, Birchwood Science Park, Warrington, Cheshire WA3 7PB, UK; Applied Biosystems Inc., 850 Lincoln Center Drive, Foster City, CA 94404, USA.

Multipurpose gas-, liquid-, solid-phase protein sequencer marketed by Knauer GmbH, Hegauer Weg 38, D-1000, Berlin 37, FRG.

C1 4000 Protein Sequencer, Chelsea Instruments Ltd, distributed in UK by Biotech Instruments Ltd, Luton.

Jaytee 8710 Protein Sequencer, Jaytee Biosciences, Reculver Road, Herne Bay, Kent, UK.

Pro-Sequencer, Milligen, Peterborough Road, Harrow, UK.

Porton Protein Sequencer, Porton Instruments, Balboa Boulevard, Encino, CA, USA.

7. REFERENCES

1. Edman,P. and Begg,G. (1967) *Eur. J. Biochem.*, **1**, 80.
2. Hewick,R.M., Hunkapiller,M.W., Hood,L.E. and Dreyer,W.J. (1981) *J. Biol. Chem.*, **256**, 7990.
3. Wittmann-Liebold,B. (1983) In *Modern Methods of Protein Chemistry*. Tschesche, H. (ed.), Walter de Gruyter, Berlin, p. 229.
4. Reimann,F. and Wittmann-Liebold,B. (1986) *Methods in Protein Sequence Analysis*. Seattle, Abstract B56.
5. Vandekerckhove,J., Bauw,G., Pupe,M., VanDamme,J. and VanMontagu,M. (1985) *Eur. J. Biochem.*, **152**, 9.
6. Hunkapillar,M.W. (1983) In *Methods in Enzymology*. Hirs,C.H.W. and Timasheff,S.N. (eds), Academic Press, New York, Vol. 91, p. 227.
7. Hansen,J.N. (1977) *Anal. Biochem.*, **76**, 37.
8. Friedman,M., Krull,L.H. and Cavins,J.F. (1970) *J. Biol. Chem.*, **245**, 3868.
9. Amons,R. (1987) *FEBS Lett.*, **212**, 68.
10. Brauer,A.W., Oman,C.L. and Margolies,M.N. (1984) *Anal. Biochem.*, **137**, 134.
11. Lottspeich,F. (1980) *Hoppe-Seyler's Z. Physiol. Chem.*, **361**, 1829.
12. Ashman,K. and Wittmann-Liebold,B. (1985) *FEBS Lett.*, **190**, 129.
13. Hunkapiller,M.W. (1985) ABI User Bulletin No. 14, Applied Biosystems.
14. Aitken,A., Bilham,T. and Cohen,P. (1982) *Eur. J. Biochem.*, **126**, 235.
15. Aitken,A., Bilham,T., Cohen,P., Aswad,D. and Greengard,P. (1982) *J. Biol. Chem.*, **256**, 3501.
16. Cook,K.G., Bradford,A.P., Yeaman,S.J., Aitken,A., Fearnley,I.M. and Walker,J.E. (1984) *Eur. J. Biochem.*, **145**, 587.
17. Annan,W.D., Manson,W. and Nimmo,J.A. (1982) *Anal. Biochem.*, **121**, 62.
18. Richardson,W.S., Munksgaard,E.C. and Bulter,W.T. (1978) *J. Biol. Chem.*, **253**, 8042.
19. Downs,F., Peterson,C., Murty,V.L.N. and Pigman (1977) *Int. J. Peptide Protein Res.*, **10**, 315.
20. Wang,Y., Bell,A.W., Hermodsen,M.A. and Roach,P.J. (1986) *Methods in Protein Sequence Analysis*. Abstract D51.
21. Meyer,H.E., Hoffmann-Posorske,E., Korte,H. and Heilmeyer,L.M.G.,Jr. (1986) *FEBS Lett.*, **204**, 61.
22. Holmes,C.F.B. (1987) *FEBS Lett.*, in press.

CHAPTER 5

Sequencing by mass spectrometry

KLAUS BIEMANN

1. INTRODUCTION

Over the past few years, mass spectrometry (MS) has emerged as an important technique for the determination of the primary structure of proteins. It is particularly useful if the protein is modified, either in the course of natural post-translational events or by chemical reactions involving new covalent bonds. These developments have become possible because of a new ionization method described in 1981 by Barber (1) and termed fast atom bombardment (FAB) because it involves the irradiation of the compound of interest in a solvent such as glycerol, with a beam of atoms (Ar or Xe) in the ion source of a mass spectrometer. The process leads to the ejection of secondary ions from the glycerol solution, among them $(M + H)^+$ ions where M stands for the solute molecule of interest, for example a peptide. This ionization method has eliminated the previous major limitation of MS, namely, that the compound of interest had to be vapourized without thermal decomposition. With FAB it is now possible to ionize large, polar molecules such as peptides and small proteins and the implications for peptide and protein chemistry have been significant (2). The capability of determining the molecular weight of a polypeptide to within one mass unit often suffices to answer particular questions. Structural information (sequence, modifications) can be deduced from characteristic species formed by fragmentation of the protonated molecular ion, $(M + H)^+$, either spontaneously or upon collision with a neutral gas. In the former case, the fragment ions appear as peaks at the corresponding mass in the conventional FAB spectrum (*Figure 1*). If collision is required to induce enhanced fragmentation, a tandem mass spectrometer has to be used, which allows the mass analysis of the fragment ions in the second of the two mass spectrometers (see Section 3.2.2).

Thus, mass spectrometry represents an entirely different approach to Edman degradation (see Chapters 2–4) which makes the two methodologies complementary in the analysis of peptide structure.

2. METHODOLOGY

As mentioned earlier, FAB ionization involves the bombardment of a solution of the peptide by an atom beam, usually of 5–10 kV kinetic energy. [It should be noted that charged particles, such as Cs^+ are equally or more effective than neutral atoms and this variant of secondary ion mass spectrometry is sometimes referred to as 'liquid SIMS' (3). For the sake of simplicity in this chapter, however, the generic term FAB–MS is used for either process.] The choice of the solvent (or matrix) is important. It has to have a very low vapour pressure to remain for 2–15 min on the sample probe, which

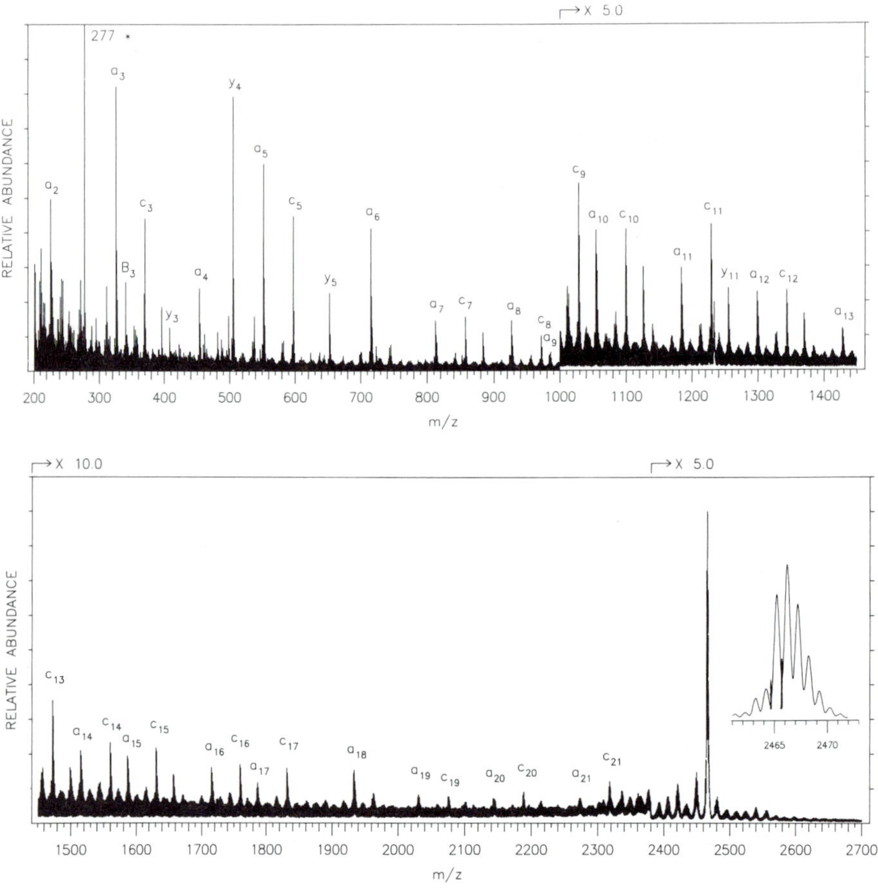

Figure 1. FAB mass spectrum of the peptide Arg-Pro-Val-Lys-Val-Trp-Pro-Asn-Gly-Ala-Glu-Ser-Ala-Glu-Ala-Phe-Pro-Leu-Glu-Phe (segment 18–39 of ACTH). For the designation a_n and c_n, see *Scheme 1*. The inset is an expansion of the $(M + H)^+$ region. The peak within the vertical lines (m/z 2465.2 ± 0.5) was selected for collision and the resulting CID spectrum is shown in *Figure 10*. (Reproduced with permission of John Wiley & Sons Ltd.)

is inserted into the ion source of the mass spectrometer where the pressure is about 10^{-6} torr. It has also to be a good solvent for the compound to be analysed. Liquid polyhydroxy compounds fulfill both requirements for peptides and glycerol has emerged as the most widely used matrix. Others which have been used include a mixture of glycerol with thioglycerol or an eutectic mixture of dithiothreitol and dithioerythritol (5:1). The thiol groups are thought to facilitate protonation of the peptide molecules but may lead to slow reduction of any disulphide bonds present.

2.1 Sample preparation

The sample is usually prepared by dissolving the peptide in 30% aqueous acetic acid at a concentration of about 0.5–1 μg/μl which is then mixed with an equal volume of the matrix. A portion (0.5–1 μl) of this mixture is then placed on the tip of a suitably shaped sample probe which is inserted into the ion source of the mass spectrometer via a vacuum lock. During evacuation in the lock most of the water and acetic acid evaporate while the matrix remains with the sample. Where amounts of peptide are limiting (1 μmol or less) great care should be taken in these manipulations. To minimize losses the peptides should be concentrated in a vacuum centrifuge and evaporation to dryness should be avoided whenever possible.

Particular attention has to be paid to the elimination of inorganic salts from the sample because they lead to the appearance of adducts with the peptide, such as $(M + Na)^+$ or $(M + K)^+$ ions, 22 and 38 daltons heavier respectively than $(M + H)^+$. While such cationated species can be used to confirm the correct assignment of M, they also reduce the $(M + H)^+$ signal, thus lowering sensitivity and complicating the spectrum. However, the most detrimental effect of salt contamination is the tendency of glycerol to form stable clusters of $(glycerol)_nNa^+$ or K^+ which often compete with $(M + H)^+$ ion formation to the point that the latter becomes undetectable. For these reasons, ammonium bicarbonate, pyridinium acetate and similar organic salts are generally used as buffers. Contaminating salts can be removed by reverse-phase high-performance liquid chromatography (HPLC) (making sure the apparatus is free of inorganic salts) or by using a Sep-Pak™ cartridge. The salts are washed off with water and the peptide eluted with methanol or acetonitrile.

2.2 Choice of mass spectrometers

In principle, any type of mass spectrometer system is suitable but for practical purposes double focusing magnetic deflection instruments are most widely used for structural studies of peptides. Quadrupole instruments are less popular while time-of-flight spectrometers have a special application (see Section 3.1.5). The magnetic instruments have the advantage of the best combination of mass range, resolution and sensitivity, at least in the region up to 10 000 daltons, which covers the usual range of peptides produced by enzymic or chemical cleavage of proteins. For tandem mass spectrometry (see Section 3.2.2), combinations of two magnetic instruments, three quadrupoles (the middle one representing the collision region), or combinations thereof (so called hybrid instruments) are employed.

3. APPLICATIONS

3.1 Structure problems requiring only molecular weight information

There are many questions that can be simply answered by the determination of the molecular weight of one or more peptides with an accuracy of the mass measurement to better than ± 0.5 dalton. This permits the reliable differentiation of two peptides that differ in mass by only one dalton, an important point because a number of amino acids differ by only that mass (see *Table 1*). It should be noted that double focusing magnetic deflection mass spectrometers are able to measure the mass of an ion with a much higher accuracy (down to an error of a few p.p.m.) which allows the differentiation of ions of the same nominal mass but different elemental composition. On rare occasions, such measurements are useful to distinguish between peptides of different composition (generally below mol. wt 2000).

Table 1. Residue masses of common amino acids.

Amino acid			*Amino acid*		
Three letter	*Single letter*	*Residue mass*[a]	*Three letter*	*Single letter*	*Residue mass*[a]
Gly	G	57.02	Asp	D	115.03
Ala	A	71.04	Lys	K	128.09
Ser	S	87.03	Gln	Q	128.06
Pro	P	97.05	Glu	E	129.04
Val	V	99.07	Met	M	131.04
Thr	T	101.05	His	H	137.07
Cys	C	103.01	Phe	F	147.07
Ile	I	113.08	Arg	R	156.10
Leu	L	113.08	Tyr	Y	163.06
Asn	N	114.04	Trp	W	186.08

[a]Mass of -NH-CHR-CO- rounded to the nearest 0.01 dalton.

3.1.1 *Confirmation or correction of the DNA sequence of a gene coding for a protein*

Frequently the primary structure of a protein has been deduced from the base sequence of the gene coding for the protein. The correctness of that sequence can be easily checked by determining the molecular weights of the peptides produced by proteolytic digestion of the protein (4). In this case, a limited number of specific peptides is required and trypsin is often the best enzyme for this purpose. The approach then taken is to match the peptide molecular weights deduced from the mass of the $(M + H)^+$ ions identified by FAB–MS to those anticipated from the tryptic cleavage sites present in the hypothetical protein sequence.

It is an indication of errors in the DNA sequence if one or more of the experimentally determined molecular weights do not correspond to any of the predicted ones. Such mismatches can be caused either by the misidentification of a nucleotide base, which may change the base triplet to the codon for a different amino acid, or by missing a base (deletion) or by inserting a fictitious base (insertion) during the interpretation of the sequencing gel. The last two cases lead to 'frame shifts' from the proper ('open', that is, free of stop codons) reading frame into one of the two other possible ways of reading triplets in a base sequence. Since either error would eventually lead to an early stop codon, it escapes detection during the DNA sequencing only if there is a pair of such errors (deletion, insertion) which causes the deduced amino acid sequence to shift into the other reading frame for a short distance. Such a case is shown in *Figure 2*, where the peptides found matched reading frame 1 from the N terminus to amino acid 324 and again from amino acid 387 to the C terminus. There were two peptides that had molecular weights predicted for reading frame 3 in that region, indicating a deletion error between positions 324 and 339 and an insertion error between positions 360 and 387.

A mismatched pair differing by the difference in the molecular weights of two amino acids indicates a misidentified base in the region where that peptide is located in the hypothetical amino acid sequence. An example is shown in *Figure 3*. The molecular weights for the tryptic peptides predicted from that region of glutamine-tRNA synthetase

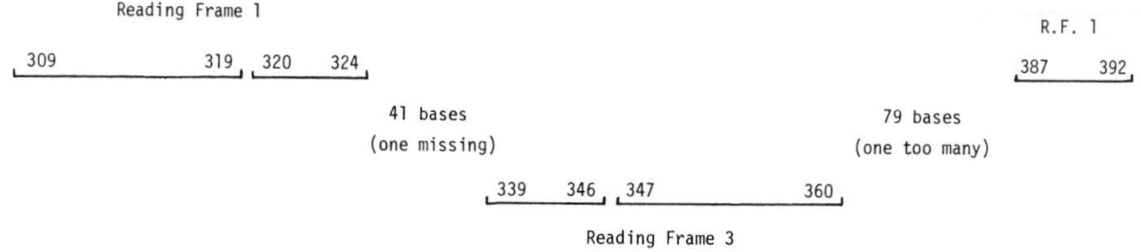

Figure 2. Detection of deletion and insertion errors.

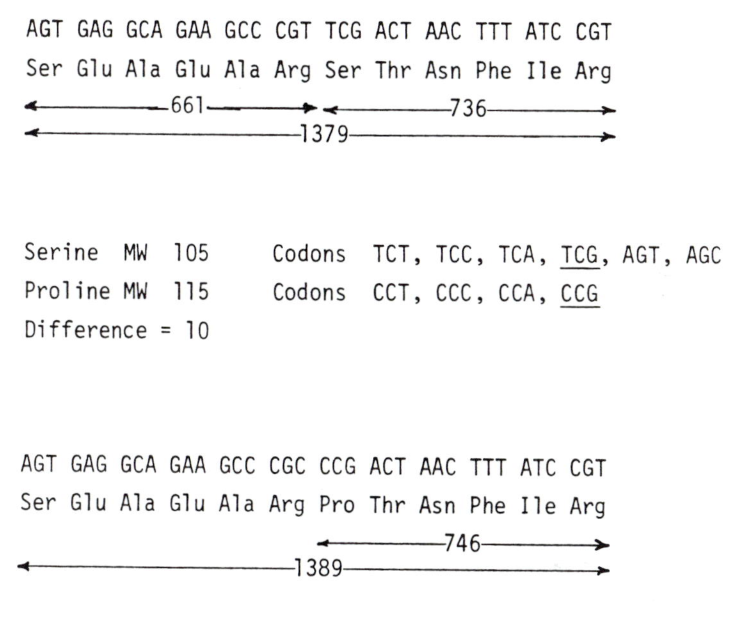

Figure 3. Detection of a mis-identified base in a DNA sequence.

(from *Escherichia coli*) which encompasses positions 1–12 would be 661 and 736, respectively. None of the peptides from the tryptic digest of this protein fit these expectations but there were two $(M + H)^+$ ions at m/z 747 and 1390 in the FAB mass spectra, indicating peptides of molecular weights 746 and 1389. The former is 10 daltons heavier than 736, indicating the presence of proline in place of serine. For this to be possible, the latter must be part of the predicted sequence (which it is) and the respective codons should differ by only one base, because the chance of two or three consecutive errors is unlikely. In this case, changing the thymine (T) in the serine codon to a cytosine (C) generates proline in this position. This also explains the formation of the other peptide of molecular weight 1389 because the Arg–Pro bond is quite resistant to trypsin and only cleaves slowly (see Chapter 2).

3.1.2 *Modified sequences*

The same strategy can be employed for checking the correctness of the base sequence of a gene chemically modified to produce an altered amino acid sequence or to identify the type and site of a naturally occurring mutation. In all these cases, it is again a matter of comparing the molecular weights of peptides determined by FAB−MS in a chosen specific digest with those expected.

3.1.3 *Experimental conditions*

As already stated, highly specific cleavage reactions are desirable to generate a relatively simple mixture of peptides and to make it possible to accurately predict the molecular weight of the expected fragments. Trypsin is the enzyme of choice because of its specificity for arginine and lysine which are amino acids that occur at moderate frequency in most proteins. One can therefore generate a limited number of peptides of intermediate length, rarely containing more than 25 amino acids and thus easily within the reach of FAB−MS techniques. Integral membrane proteins present greater problems and cleavage with CNBr (see Section 4.2.6, Chapter 2) may be more appropriate. Even then separation is a major problem (see ref 20, Chapter 6).

If only molecular weight information is required, a few nanomoles of protein are digested with trypsin at a 100:1 (w/w) substrate/enzyme ratio in 0.1 M NH_4HCO_3, 0.1 mM $CaCl_2$ (pH 8.0−8.5) at 37°C. It is advisable to follow the progress of the proteolysis by analysing a small portion of the digest by HPLC (with detection at 210−214 nm) until the protein has degraded to the point where the chromatographic pattern has stabilized. The major portion of the digest is then subjected to a fractionation step such as a reverse-phase HPLC using a solvent gradient from 100% water to 100% acetonitrile (both phases containing 0.1% trifluoroacetic acid). There is no need to attempt complete separation of all the peaks because one of the major advantages

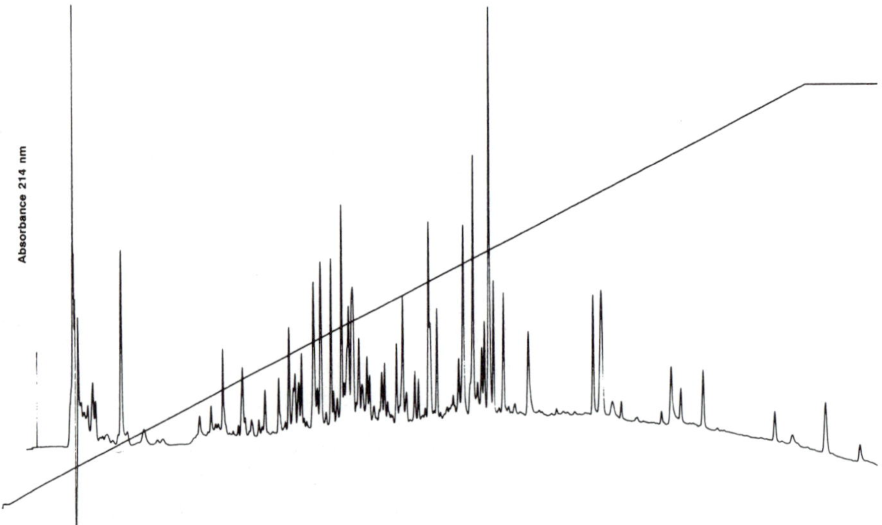

Figure 4. Partial separation of a tryptic digest of valyl-tRNA synthetase from yeast by HPLC.

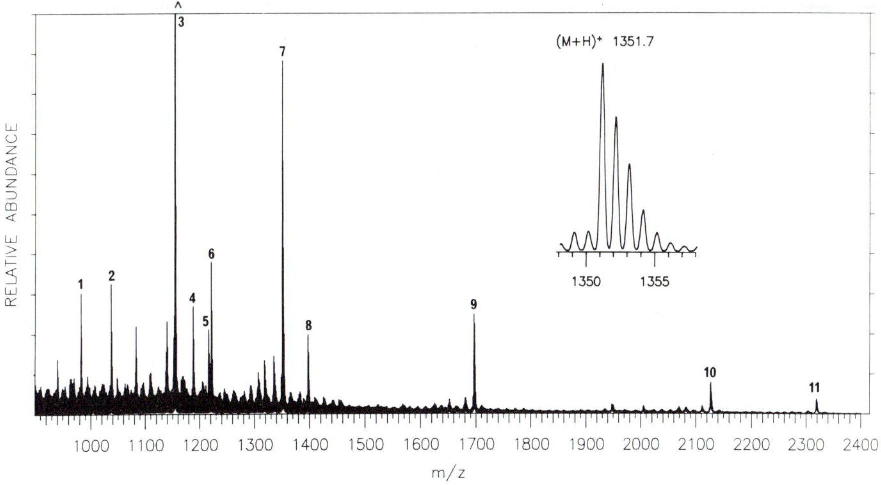

Figure 5. Mass spectrum of a fraction collected from the HPLC profile shown in *Figure 4*. (Reproduced with permission of the American Association of Advanced Science.)

of FAB−MS is its capability to analyse mixtures of peptides. Depending on the size of the protein and the expected complexity of the digest, 5−50 fractions may be taken. For example, during the chromatography shown in *Figure 4*, which represents a 2−3 nmol aliquot from a 10 nmol tryptic digest of cytoplasmic valyl-tRNA synthetase from yeast, 27 fractions were collected (5). Each of them was evaporated to dryness, the residue treated as outlined in Section 2.1 and subjected to FAB−MS. A total of 80 peptides was detected in these fractions. The spectrum obtained with fraction 10 is shown in *Figure 5*. It clearly indicates the presence of at least 11 different peptides (some of the minor peaks may be due to low yield peptides or to fragments of the more abundant ones, but it is better to be conservative and utilize only the major signals). The data are recorded as mass spectral peak profiles and, therefore, can be plotted and expanded for any region. This is shown in the inset for the cluster of peaks surrounding m/z 1351.7, which is the ^{12}C isotope component of this particular $(M + H)^+$ ion and indicates that the peptide has a molecular weight of 1350.7. All masses are automatically assigned by the data system of the mass spectrometer after proper calibration with the $(CsI)_nCs^+$ cluster ion, easily formed by FAB ionization of a thin layer of solid caesium iodide.

3.1.4 *Post-translational modifications*

Most proteins are subject to enzymic processing after biosynthesis. Examples of such modifications include the clipping off of amino acids at the N or C termini by peptidases, acylation of the amino terminus, glycosylation, phosphorylation or sulphation of certain amino acids, etc. The experiment described at the end of Section 3.1.3 was carried out to identify the N terminus of the protein and elucidate the precise modification that apparently rendered it resistant to the Edman degradation. The amino acid sequences of the N-terminal tryptic peptides predicted from the DNA sequence of the gene and the positions of the initiating methionines would be either Met-Asn-Lys and Met-Asn-

Met	Asn	Lys	Trp	Leu	Asn	Thr	Leu	Ser	Lys	Thr	Phe	Thr	Phe	Arg	15
ATG	AAT	AAG	TGG	TTA	AAC	ACA	TTA	TCT	AAG	ACA	TTC	ACT	TTT	CGG	45
Leu	Leu	Asn	Cys	His	Tyr	Arg	Arg	Ser	Leu	Pro	Leu	Cys	Gln	Asn	30
CTT	TTG	AAC	TGT	CAT	TAT	AGG	CGA	TCA	TTA	CCA	CTT	TGT	CAA	AAC	90
Phe	Ser	Leu	Lys	Lys	Ser	Leu	Thr	His	Asn	Gln	Val	Arg	Phe	Phe	45
TTT	TCT	CTG	AAG	AAG	TCG	TTA	ACT	CAT	AAT	CAA	GTC	AGG	TTC	TTT	135
Lys	Met	Ser	Asp	Leu	Asp	Asn	Leu	Pro	Pro	Val	Asp	Pro	Lys	Thr	60
AAA	ATG	AGC	GAT	CTT	GAT	AAT	TTG	CCT	CCA	GTT	GAC	CCA	AAG	ACT	180
...	...	...	...	...	...	...	...	...	...	...	...	...	...	...	....
...	...	...	...	...	...	...	...	...	...	...	...	...	...	...	....
Ile	Glu	Asn	Leu	Lys	Arg	Leu	Lys	Leu	1104						
AAT	GAA	AAC	TTG	AAG	CGT	TTG	AAA	TTG	3312						

Figure 6. Partial sequence of the gene that encodes valyl-tRNA synthetase from *Saccharomyces cerevisiae*, as determined by F.Fasiolo *et al.*, unpublished data. The two putative initiation translation sites are underlined. (Reproduced with permission of the American Association of Advanced Science.)

Lys-Trp-Leu-Asn-Thr-Leu-Ser-Lys, or Met-Ser-Asp-Leu-Asp-Asn-Leu-Pro-Pro-Val-Asp-Pro-Lys depending on whether initiation takes place at the first or second methionine of the predicted sequence shown in *Figure 6.*

The species of $(M + H)^+$ 1351.7 did not match the molecular weight of any of the predicted peptides from the protein. However, the value is close to 1440.7, the $(M + H)^+$ ion predicted for the third sequence above. Simple arithmetic reveals that the actual peptide obtained lacks the terminal methionine, which was post-translationally removed (loss of 131 daltons), and that the newly formed N terminus was acetylated (gain of 42 daltons). It should be emphasized that this conclusion could be drawn from a single FAB mass spectrum of an HPLC fraction that contained at least 11 peptides. The molecular weights of the other 79 peptides identified in the 27 HPLC fractions of the tryptic digest matched those predicted and corresponded to approximately 80% of the protein. The same experiment, therefore, also indicated that the DNA sequence of the gene was correct.

3.1.5 *Molecular weight determination of small proteins*

(i) *By FAB–MS.* Even quite large polypeptides can be ionized by FAB. The classical example is insulin which has become a mass spectrometric standard test compound in the high mass range. The $(M + H)^+$ regions for porcine and human insulins (which differ by Ala versus Thr at the C terminus of the B-chain) determined with a magnetic deflection mass spectrometer (JEOL HX110) are shown in *Figure 7.* Because of the large number of carbon atoms ($C_{256}H_{381}N_{65}O_{76}S_6$ and $C_{257}H_{383}N_{65}O_{77}S_6$, respectively), the $(M + H)^+$ ions made up of only ^{12}C (m/z 5774.6 and 5804.6) are of much lower abundance than those containing three ^{13}C atoms and which are three daltons heavier. Thus, while it is possible to determine the mass of the individual isotopic components with an accuracy of a few tenths of a dalton, it is not easy to tell which one is the 'monoisotopic' (^{12}C only) peak if one deals with a compound of unknown molecular weight. A mis-assignment by one or even two daltons could result. For this reason, and also to trade sensitivity for resolution, molecular weight determinations in this mass range and higher are usually carried out at much lower resolving power.

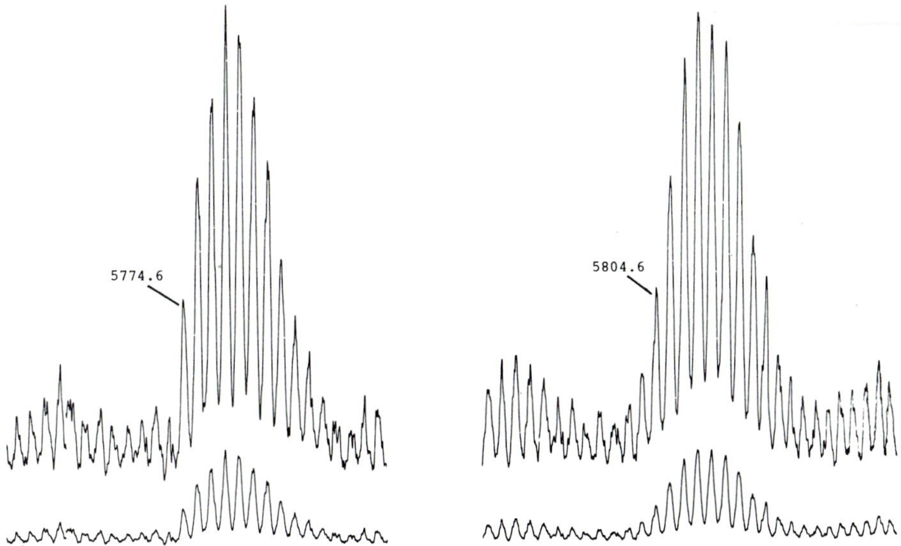

Figure 7. The molecular ion region of the FAB spectra of porcine and human insulins, respectively.

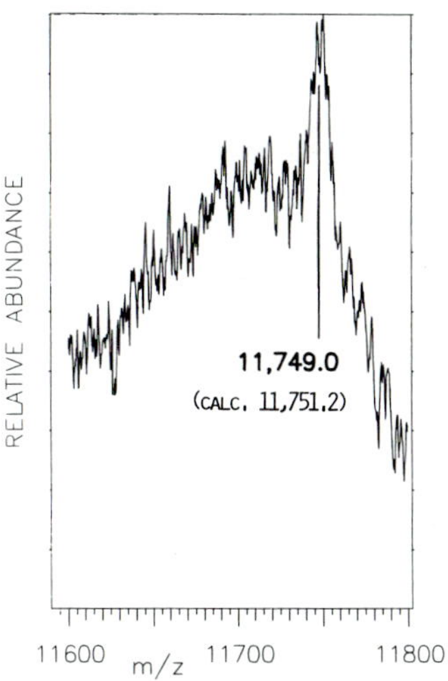

Figure 8. Unresolved isotopic multiplet of the $(M + H)^+$ ion of thioredoxin from *Chromatium vinosum*. The mass value shown corresponds to the centre of the peak. (Reproduced with permission of the American Chemical Society.)

Under these conditions, the isotopic cluster collapses to a relatively smooth peak envelope, the centroid of which represents the average mass of the isotopic components and thus the average molecular weight. This can be calculated for a given elemental composition by adding the average atomic weights of the elements present (i.e. 12.011 rather than 12.000 for carbon, etc.). For the $(M + H)^+$ ions of porcine and human

$$\left[H_2N{-}CH(R){-}CO{-}\left(NH{-}CH(R){-}CO\right)_x{-}NH{-}CH(R){-}COOH + H \right]^+$$

I

$$H{-}\left(HN{-}CH(R){-}CO\right)_{n\text{-}1}{-}\overset{+}{N}H{=}CH(R_n)$$

$\underline{a_n}$ $(\Sigma_n\text{-}27)$

$$^{+}CO{-}NH{-}CH(R_n){-}CO{-}\left(NH{-}CH(R){-}CO\right)_{n\text{-}1}OH$$

x_n $(\Sigma_n{+}45)$

$$H{-}\left(HN{-}CH(R){-}CO\right)_{n\text{-}1}{-}NH{-}CH(R_n){-}C{\equiv}O^{+}$$

$\underline{b_n}$ $(\Sigma_n{+}1)$

$$H_3\overset{+}{N}{-}CH(R_n){-}CO{-}\left(NH{-}CH(R){-}CO\right)_{n\text{-}1}OH$$

$\underline{y_n}$ $(\Sigma_n{+}19)$

$$H{-}\left(HN{-}CH(R){-}CO\right)_{n\text{-}1}{-}NH{-}CH(R_n){-}CO{-}\overset{+}{N}H_3$$

c_n $(\Sigma_n{+}18)$

$$^{+}CH(R_n){-}CO{-}\left(NH{-}CH(R){-}CO\right)_{n\text{-}1}OH$$

z_n $(\Sigma_n{+}2)$

$$\left[H{-}\left(NH{-}CH(R){-}CO\right)_{n\text{-}1}{-}NH\right]H^{+}{-}CH{=}CHR'$$

d_n

$$R'CH{=}CH{-}CO{-}\left[\left(NH{-}CH(R){-}CO\right)_{n\text{-}1}{-}OH\right]H^{+}$$

w_n

$$HN{=}CH{-}CO{-}\left[\left(NH{-}CH(R){-}CO\right)_{n\text{-}1}{-}OH\right]H^{+}$$

v_n

$$H_2\overset{+}{N}{=}CH(R)$$

immonium ion

$$H_2N{-}CH(R_{>1}){-}\left(CO{-}NH{-}CH(R)\right)_m{-}C{\equiv}O^{+}$$

internal fragment $(\Sigma_{m+1}{+}1)$

Scheme 1. Fragments produced from protonated linear peptides (I).

insulins, these values are 5778.7 and 5808.7, respectively. *Figure 8* represents the peak envelope of a larger protein (the thioredoxin from *Chromatium vinosum*), which gave an average molecular weight of 11 749.0, within 2.1 daltons of the value calculated for the sequence determined by tandem mass spectrometry (see Section 3.2.3).

(ii) *By plasma desorption mass spectrometry (PD–MS).* For molecular weight determinations of even larger proteins, ionization by MeV particles, generated by the radioactive decay of ^{252}Cf, is more efficient than FAB. Molecular weights of up to 30 000 daltons have been measured with an accuracy of $\pm 0.1-0.2\%$, which is still within the mass of one amino acid (6). At present, it is fair to say that below mass 10 000, FAB in combination with magnetic instruments gives better results while above this mass, PD–MS time-of-flight mass spectrometers may be more successful.

3.2 **Problems that require sequence information**

3.2.1 *Sequence data obtained from conventional FAB mass spectra*

The spectrum shown in *Figure 1* reveals that some fragmentation takes place upon FAB. The peaks labelled a_n and c_n differ by the residue mass (see *Table 1*) of the consecutive amino acids of this peptide and from this data the sequence can be deduced (for an explanation for the notation see Section 3.2.2 and *Scheme 1*). To obtain such an interpretable spectrum, a relatively large sample (2–3 nmol) must be used and it must be a single component because, for a mixture, the contributions of the various components could not be reliably sorted out. It should also be noted that the particular peptide giving rise to the spectrum shown in *Figure 1* represents a very favourable example.

Normally, there is much less fragmentation and chemical or enzymic methods can be used to make up for this lack of sequence information. Most often stepwise, consecutive removal of N-terminal amino acids by manual Edman degradation, followed by re-determination of the molecular weights of the shortened peptides provides sequence information. This strategy can also be used with simple mixtures by correlating each consecutive set of $(M + H)^+$ ions which always must be lower by one of the values of *Table 1*. After each step, an aliquot is taken for FAB–MS and the remainder put through another Edman cycle. In general, up to five consecutive steps can be carried out before the mass spectra become too weak and noisy to be reliably interpretable. In the very first step, information about the number of lysine residues in the peptide is obtained because each one combines via the ϵ-amino group with one molecule of phenylisothiocyanate thereby increasing the molecular weight by 135 daltons for each lysine in the peptide.

A similar approach utilizes carboxypeptidases to obtain C-terminal sequence information by following the appearance of shorter and shorter peptides in the FAB spectra taken at defined intervals (7).

3.2.2 *Sequencing by tandem mass spectrometry*

The energy imparted onto an $(M + H)^+$ ion by FAB is not sufficient to cause extensive fragmentation. In order to overcome this problem, additional energy can be provided by collision with an inert atom, such as helium. This is best accomplished in a tandem mass spectrometer where FAB ionization takes place in the first mass spectrometer (MS-1, see *Figure 9*), which not only provides the normal spectrum of the sample but also permits the selection of any one ion for transmission into the collision cell containing helium at approximately 10^{-3} torr. The fragment ions produced upon the collision-induced decomposition (CID) of the precursor ion are then mass separated in the second

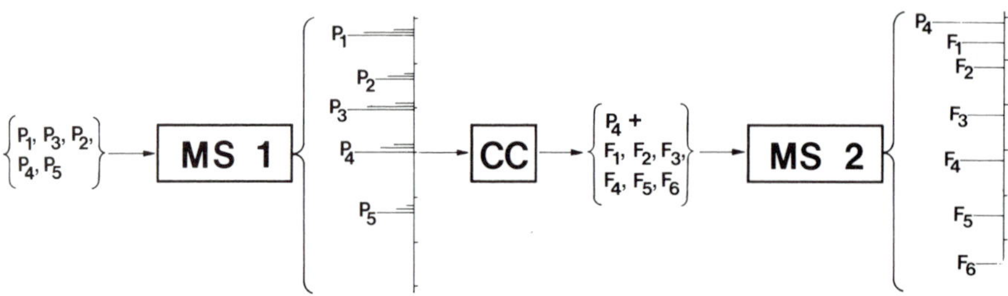

Figure 9. Principle of tandem mass spectrometry. The FAB mass spectrum of a mixture of five peptides are ionized in the ion source of MS-1 and the m/z values of their $(M + H)^+$ ions recorded ($P_1 - P_5$). After that, the ^{12}C-only component of each isotopic cluster is consecutively selected by MS-1 and passed through the collision cell containing approximately 10^{-3} torr of He. The collision-induced decomposition (CID) of the precursor ion and the resulting fragments $F_1 - F_6$ (product ions) are mass analysed by scanning MS-2 in the linked scan mode, where the ratio of the electric field (E_2) and magnetic field (B_2) is changed such as to keep B_2/E_2 constant. This process is repeated for each of the $(M + H)^+$ ions produced in MS-1 from the original mixture. Note that the ions in the MS-1 scan are polyisotopic while those in the MS-2 scan are monoisotopic. (Reproduced with permission of John Wiley & Sons Ltd.)

mass spectrometer (MS-2). If it is a double focusing magnetic instrument, it has to be scanned in such a manner that the fragment ions, which have lost kinetic energy proportional to the mass of the neutral fragment, are transmitted through the electric sector of MS-2. This is accomplished by changing the magnetic and electric fields at a constant ratio (a so-called B/E linked scan). In principle, this can also be done with a single mass spectrometer where the collision cell is located immediately following the ion source. The tandem arrangement of two mass spectrometers has the great advantage that a single mass, that is, the ^{12}C species of the $(M + H)^+$ ion cluster, can be selected for CID and the product ion spectrum is derived exclusively from this ion. Clearly, for mixture analysis this is very important. Furthermore, the product ion spectrum consists then of only monoisotopic ions, that is, single peaks, which means that the spectrum is very simple and, if necessary, one can sacrifice resolution for sensitivity. The CID spectrum of the same peptide represented in *Figure 1* is shown in *Figure 10*. It is dominated by fragment ions of the type a_n and d_n (see below) and the sequence can be read directly from the mass differences of the a_n series using the values in *Table 1* (for the differentiation of leucine and isoleucine, and lysine and glutamine, see below and Section 3.2.3).

The above discussion centred around a tandem mass spectrometer consisting of two double focusing magnetic spectrometers, a combination which provides the best resolution and sensitivity for both the precursor ions in MS-1 and product ions in MS-2 (8). Another arrangement is a triple quadrupole mass spectrometer in which the first quadrupole is MS-1, the second serves as the collision region and the last as MS-2 (9). The major difference is in the collision process which in the second arrangement involves multiple collisions at low kinetic energy (10 – 100 eV). In magnetic instruments the precursor ions usually have about 10 keV kinetic energy and suffer only a single collision while passing through the relatively short (1 – 2 cm) collision cell. Some of the fragmentation processes therefore may differ. A limitation of the triple quadrupole is their lower resolution and mass range when compared with magnetic spectrometers.

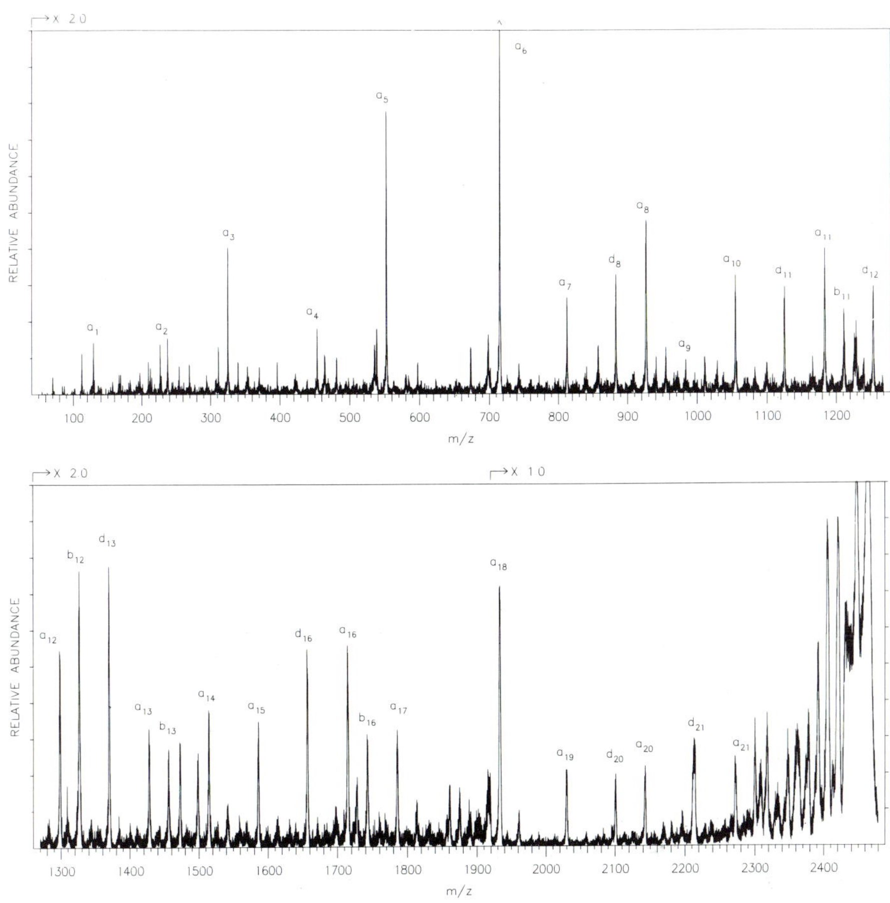

Figure 10. The CID spectrum of the ^{12}C-only component of the $(M + H)^+$ ion shown in *Figure 1*. For the designation a_n, b_n and d_n, see *Schemes 1* and *2*, respectively. (Reproduced with permission of John Wiley & Sons Ltd.)

There are also so-called hybrid tandem mass spectrometers where MS-1 is a magnetic instrument and MS-2 a quadrupole.

(i) *Fragmentation of peptides upon CID. Scheme 1* summarizes the fragment ions produced upon high energy (keV) collision of a $(M + H)^+$ ion with a neutral gas such as helium. The ions of type a_n, b_n, c_n, x_n, y_n and z_n result from cleavage of a peptide bond and are quite independent of the nature of the amino acid at that particular site. Therefore, they can appear as extended continuous series of ions (particularly a_n, b_n and y_n) from which the sequence may be read in either one or both directions. It may be noted that c_n ions are frequently observed in normal FAB spectra (see *Figure 1*), but rarely in high energy CID spectra (*Figure 10*).

The ions of type d_n, v_n and w_n involve fragmentation of the side chain at or near the point of backbone cleavage and thus are formed only from certain amino acids.

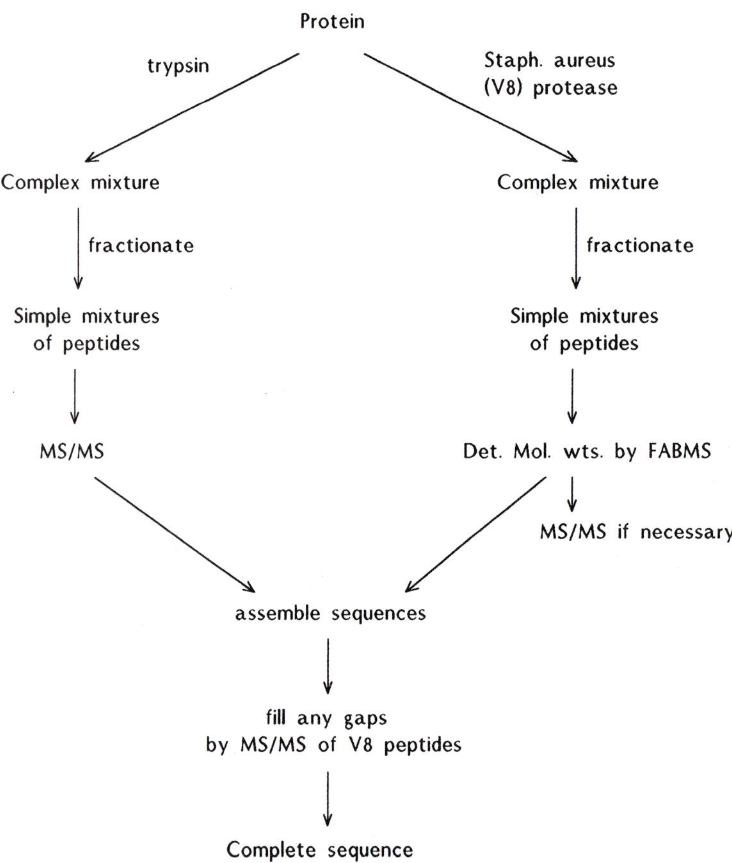

Scheme 2. Strategy for the determination of the amino acid sequence of proteins by tandem mass spectrometry.

The d_n and w_n ions allow the differentiation of leucine and isoleucine because the substituent of the β-carbon is retained.

The immonium ions result from cleavage of a bond at either side of an amino acid. Their mass indicates the presence of certain amino acids (see *Figure 11a*). Larger internal fragments represent sub-sequences and are often pronounced when proline is present. They are generally labelled using the single letter code for amino acids.

3.2.3 *Applications of tandem mass spectrometry*

(i) *N-blocked peptides.* The Edman degradation procedure allows the direct determination of the N-terminal sequence of an intact protein (which cannot be done by MS) for tens of steps if there is a basic, primary or secondary amino group at the N terminus. However, this is not possible if the amino terminus is acetylated and in such situations the MS approach is invaluable. The acyl group merely increases the mass of the N-terminal amino acid, that is, shifts the molecular weight of all peaks due to ions containing the N terminus. An example is shown in *Figure 11a* which also illustrates the information that can be derived from the immonium ions in the low mass region,

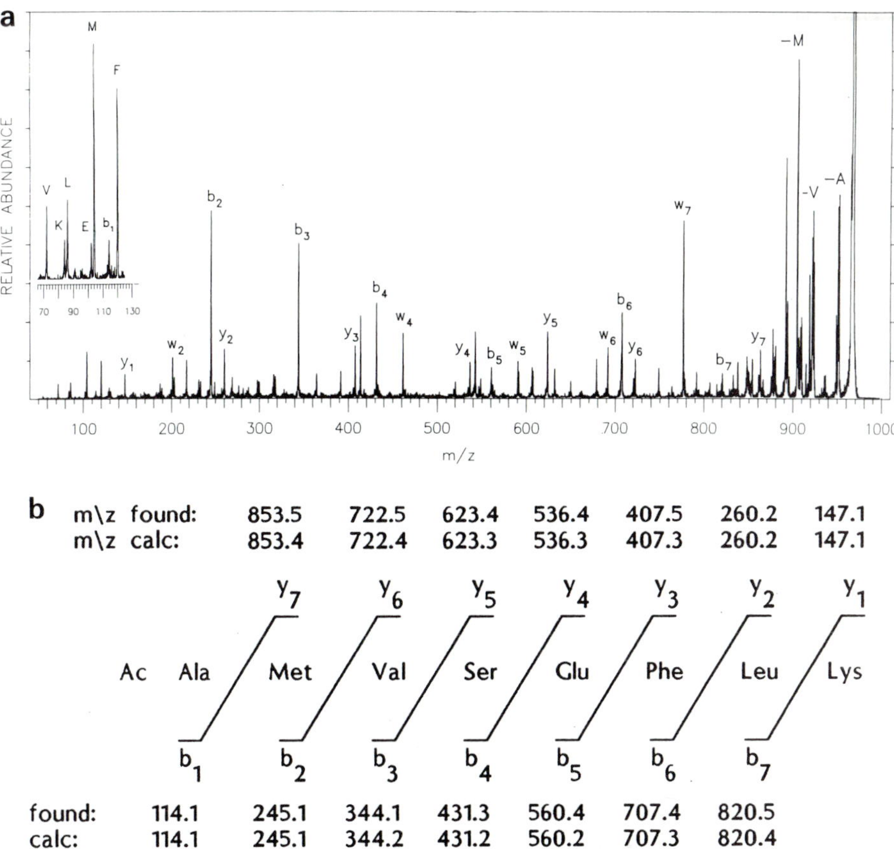

Figure 11. (**a**) The CID spectrum of the $(M + H)^+$ ion of the N-terminal tryptic peptide of lipocortin I. **Insert**: low mass region expanded. (**b**) Amino acid sequence as derived from mass spectral data. The y_n and b_n fragment ions are shown including their calculated and experimentally determined masses.

namely, the presence of the amino acids Val, Lys (the immonium ion cyclyzes by elimination of NH_3), Leu (or Ile), Glu, Met and Phe at m/z 72, 84, 86, 102, 104 and 120. The m/z values for the b_n and y_n series are listed in *Figure 11b* which also gives an indication of the accuracy of the mass assignment.

(ii) *Other modifications.* Just as the acetylation of alanine shifts the b_1 ion to m/z 114 in the above example, any change in mass of an amino acid will shift the normal sequence ions containing this amino acid. Examples are:

(a) the replacement of one amino acid by another through mutation (10);
(b) phosphorylation (11) or sulphation of hydroxy amino acids;
(c) the attachment of *N*-acetylglucosamine to asparagine which increases its mass by 203 daltons (12);
(d) the formation of a new covalent bond to an external molecule in a photoaffinity labelling experiment (13); or
(e) many other chemical or enzymic events.

Once the primary sequence of a protein is known, disulphide bridges can be assigned by determination of the molecular weights of appropriate peptides before and after reduction of their disulphide bonds (14).

Here again the complementarity with or advantage over the Edman method is evident, because some of these modifications are either lost under the Edman reaction conditions, or the products cannot be extracted or identified by HPLC.

(iii) *Determination of the primary structure of proteins.* Sequencing by tandem mass spectrometry with the presently available techniques is reliable for peptides up to about 25 amino acids in length. This is just about the size of the peptides that are usually generated upon digestion of an 'average' water-soluble protein with trypsin, or *Staphylococcus aureus* (strain V8) protease. Thus, the simple strategy outlined in *Scheme 2* can be followed to determine the primary structure of a protein.

At this point one could, in principle, take the entire digest, after removal of the buffer salts, and subject it to FAB–MS for the determination of the molecular weights of all peptides present, followed by CID of each $(M + H)^+$ ion. However, it is advisable to separate this relatively complex mixture by HPLC into fractions containing a few (1–5) peptides each. This step not only removes any salt contamination but also groups peptides of similar hydrophilicity/phobicity together, which minimizes the effect of suppression of ionization of hydrophilic peptides by hydrophobic peptides which is often observed (15). There is another practical reason for simplifying the peptide mixture to be subjected to tandem mass spectrometry. All components of a mixture are simultaneously ionized in the ion source of MS-1. However, only one $(M + H)^+$ ion at a time is subjected to CID and the MS-2 scans take 1–3 min. During this time the $(M + H)^+$ ions of all other components are wasted. After a few MS-2 scans, the sample is depleted and will have to be re-loaded. It is, therefore, in the interests of sample conservation to keep the number of peptides per fraction as low as possible.

For at least two reasons, peptides produced by tryptic digestion are most suitable for the first sequencing experiments. First, this enzyme is very specific (although care has to be taken to use highly purified enzyme, such as the preparation now commercially available from Boehringer Mannheim). Second, the position of basic amino acids influences the fragmentation of the $(M + H)^+$ ions. Since they will be at the C terminus of tryptic peptides (except for the peptide representing the original C terminus), the y_n ion series (*Scheme 1*) generally will be relatively abundant and continuous.

Interpretation of the CID mass spectra of the tryptic peptides will reveal most or all of their sequences. Remaining problems may be:

(a) the occasional inability to deduce the nature and sequence of the first or last two amino acids because the ions due to cleavage of the N- or C-terminal amino acid are often of low abundance;
(b) difficulty in the differentiation of leucine and isoleucine if the corresponding d_n or w_n ions cannot be detected; or
(c) the inability to distinguish lysine from glutamine.

The N-terminal amino acid can be identified by a single Edman step and re-determination of the molecular weight of the truncated peptide(s); this experiment then

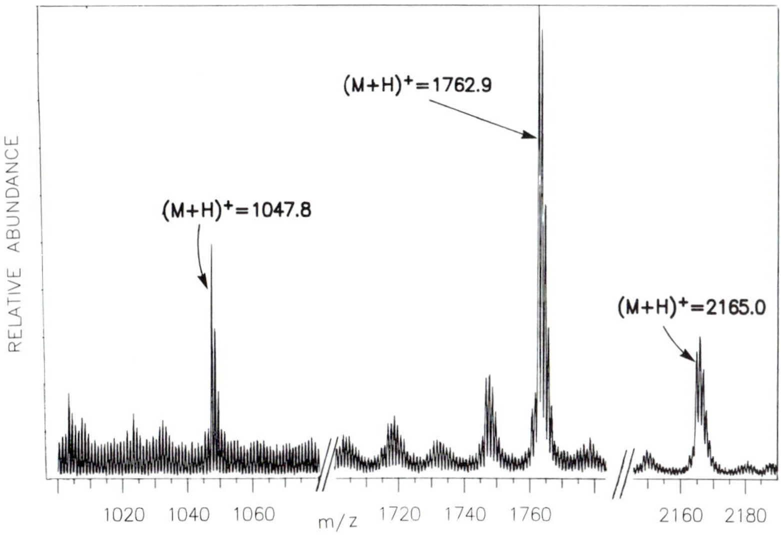

Figure 12. Sections of the FAB mass spectrum of fraction 10 of an HPLC separation of the tryptic peptides of the thioredoxin from *Chromatium vinosum*. (Reproduced with permission of John Wiley & Sons Ltd.)

also allows identification of the second amino acid, if at least one of the a_2, b_2 or y_{n-2} ions is present. Alternatively, the CID spectra of the corresponding *S. aureus* (V8) peptides can be determined. Such a digest is needed in any case to arrange the tryptic peptides in the proper order to arrive at the sequence of the intact protein. The CID spectra generally allow the determination of the remaining N- or C-terminal sequences of tryptic peptides and the differentiation of leucine and isoleucine. Because the presence and location of basic amino acids greatly influences the formation of d_n and w_n ions, leucine/isoleucine identifications which are not possible in one set of peptides (often when the amino acid is at the N or C terminus) may be possible in the other set. Finally, the distinction of lysine and glutamine is best made by treating the sample—even when still in glycerol on the FAB probe—with a small amount of phenylisothiocyanate as indicated above.

An example of the strategy outlined above is the determination of the primary structure of the thioredoxin isolated from *C. vinosum* (16). A tryptic digest of the protein was separated into 10 fractions by HPLC. Each of these fractions contained 1–3 peptides, some of them present in overlapping adjacent fractions. The FAB spectrum of fraction 10 (*Figure 12*) indicated the presence of three $(M + H)^+$ ions at m/z 1047.8, 1762.9 and 2165.0. The CID spectrum of m/z 2165.0 is shown in *Figure 13*. It reveals an almost complete sequence but has some of the shortcomings mentioned above: the first and last two amino acids are not clearly established and leucine/isoleucine is not differentiated. Since it is a tryptic peptide, the C terminus must be either lysine or arginine and the data fit only the combination carboxymethylcysteine (cys*)-lysine. The N-terminal amino acid was shown by an Edman step after chymotryptic cleavage of this peptide to be serine, which defined the N terminus as Ser-Pro. The sixth amino acid was identified as leucine based on the d_6 ion in the CID spectrum of a chymotryptic peptide (see *Figure 14*).

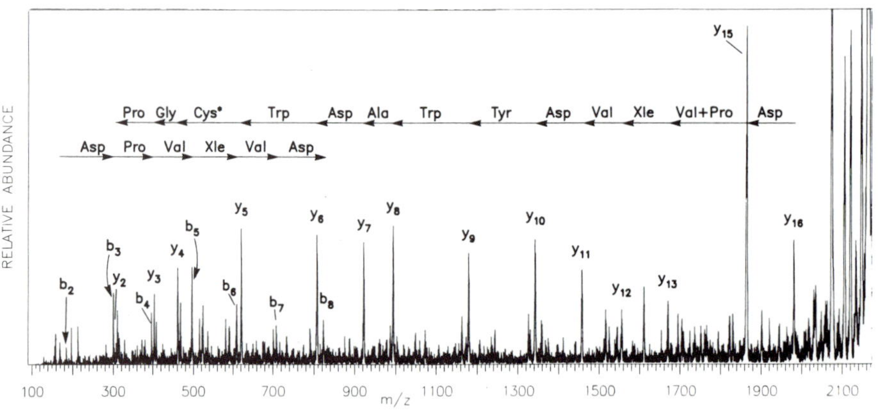

Figure 13. CID spectrum of the ^{12}C-only $(M + H)^+$ ion of m/z 2165.0 of *Figure 12*. Cys* denotes carboxymethylated cysteine. (Reproduced with permission of John Wiley & Sons Ltd.)

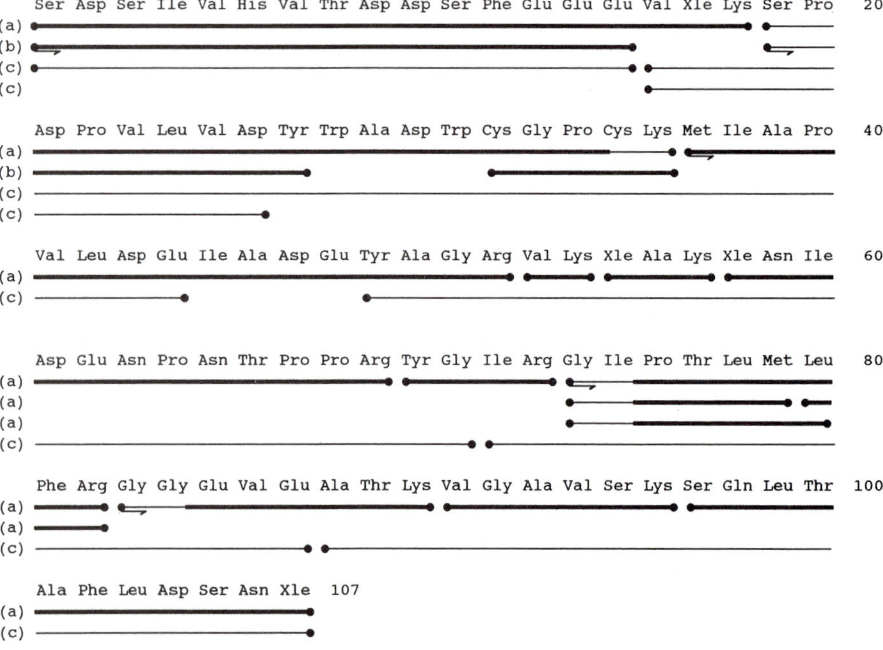

Figure 14. Amino acid sequence of the thioredoxin from *Chromatium vinosum*: (**a**) tryptic peptides (positions 74–79, 74–80 and 80–82 are due to chymotryptic cleavages); (**b**) further digestion of selected HPLC fractions from the tryptic digest with either α-chymotrypsin or *S.aureus* (strain V8) protease; (**c**) *S.aureus* (strain V8) protease peptides. Half arrows: amino acids removed by manual Edman degradation. (Reproduced with permission of John Wiley & Sons Ltd.)

The tryptic peptides were aligned as shown in *Figure 14* on the basis of the molecular weights and partial sequences of the peptides generated by cleavage of the protein with *S.aureus* (V8) protease. To check the completeness of the structure, the molecular weight of the thioredoxin was then determined and found to be within two daltons of the value calculated for this sequence (see Section 3.1.5 and *Figure 8*).

3.3 Cost effectiveness of tandem mass spectrometry for peptide sequencing

Although FAB−MS and tandem mass spectrometry may seem to be much more expensive and less sensitive than the gas-phase Edman method, the differences in sample preparation and capabilities seem to indicate this is not necessarily so and that for maximum efficiency, a well-organized protein structure laboratory should have both capabilities. The most important advantage of the mass spectrometric approach is the elimination of the need for complete separation and purification of the peptides to be sequenced. This alone partially compensates for the lower sensitivity, because less material, time and expense is required for peptide fractionation.

The speed of the mass spectrometric method is another factor saving time and resource. As described in Sections 3.2.2 and 3.2.3, the CID mass spectra of a mixture of three peptides, for example, can be obtained in a 15-min experiment, resulting in at least 30 residues of sequence. A single Edman cycle takes longer than this. In situations where there are a large number of proteins to be investigated, much more than one automatic (Edman) instrument would be required to carry out the same workload. In addition to these considerations, there is the ability of the mass spectrometer to answer certain structural questions where the Edman method is either not applicable or provides only indirect information.

4. CONCLUSION AND OUTLOOK

It is evident that over the past few years mass spectrometry has become a powerful method for the solution of various problems which arise in the course of the determination of the primary structure of peptides and proteins. This is even more important in view of the multitude of questions that arise in the study of altered products generated by recombinant technology, investigations of substrate−receptor interactions and modifications of native proteins. Further advances in mass spectrometric instrumentation and techniques, such as the recent increase in sensitivity with the use of multichannelplate detectors (17,18) or novel sample introduction techniques (19) will make mass spectrometry even more powerful. Most important is the complementarity with the Edman degradation (which until now was the only practical useful method available in this field) because the two methodologies are based on entirely different chemical and physical principles.

5. REFERENCES

1. Barber,M., Bordoli,R.S., Sedgwick,R.D. and Tyler,A.N. (1981) *J. Chem. Soc. Chem. Commun.*, 325.
2. Biemann,K. and Martin,S.A.A (1987) *Mass Spectrom. Rev.*, **6**, 1.
3. Aberth,W., Straub,K.M. and Burlingame,A.L. (1982) *Anal. Chem.*, **54**, 2029.
4. Gibson,B.W. and Biemann,K. (1984) *Proc. Natl. Acad. Sci. USA*, **81**, 1956.
5. Biemann,K. and Scoble,H.A. (1987) *Science*, **237**, 992.
6. Sundqvist,B., Kamensky,I., Hakansson,P., Kjellberg,J., Salehpour,M., Widdiyasekera,S., Fohlman,J., Peterson,P.A. and Roepstorff,P. (1984) *Biomed. Mass Spectrom.*, **11**, 242.
7. Caprioli,R.M. and Fan,T. (1986) *Biochem. Biophys. Res. Commun.*, **141**, 1058.
8. Sato,K., Asada,T., Ishihara,M., Kunihiro,F., Kammei,Y., Kubota,E., Costello,C.E., Martin,S.A., Scoble,H.A. and Biemann,K. (1987) *Anal. Chem.*, **59**, 1652.

9. Hunt,D.F., Yates,J.R., Shabanowitz,J., Winston,S. and Hauer,C.R. (1986) *Proc. Natl. Acad. Sci. USA*, **83**, 6233.
10. Katakuse,I., Ichihara,T., Nakabushi,H., Matsuo,T., Matsuda,H., Wada,Y. and Hayashi,A. (1984) *Biomed. Mass Spectrom.*, **11**, 386.
11. Petrilli,P., Pucci,P., Morris,H.R. and Addeo,F. (1986) *Biochem. Biophys. Res. Commun.*, **140**, 28.
12. Reddy,V.A., Johnson,R.S., Biemann,K., Williams,R.S., Ziegler,F.D., Trimble,R.B. and Maley,F. (1988) *J. Biol. Chem.*, **263**, 6978.
13. Brinnegar,A.C.,Cooper,G., Stevens,A., Hauer,C.R., Shabanowitz,J., Hunt,D.F. and Fox,J.E. (1988) *Proc. Natl. Acad. Sci. USA, 85*, 5927.
14. Morris,H.R. and Pucci,P. (1985) *Biochem. Biophys. Res. Commun.*, **126**, 1122.
15. Naylor,S., Findeis,A.F., Gibson,B.W. and William,D.H. (1986) *J. Am. Chem. Soc.*, **108**, 6359.
16. Johnson,R.S. and Biemann,K. (1987) *Biochemistry*, **26**, 1209.
17. Cottrell,J.S. and Evans,S. (1987) *Anal. Chem.*, **59**, 1990.
18. Hill,J.A., Martin,S.A., Biller,J.E. and Biemann,K. (1988) *Biomed. Environm. Mass Spectrom.*, **17**, 147.
19. Caprioli,R.M., Fan,T. and Cottrell,J.S. (1986) *Anal. Chem.*, **58**, 2949.
20. Findlay,J.B.C. and Evans,W.M. (eds) (1987) *Biological Membranes: A Practical Approach.* IRL Press, Oxford.

CHAPTER 6

Manual methods of protein sequencing

A.YARWOOD

1. INTRODUCTION

In these days of high technology, when sophisticated highly efficient sequencers are capable of the automated sequencing of up to 100 N-terminal residues of peptides and proteins, the manual sequencer and his methods may seem to the uninitiated to be something of an anachronism, and yet it is probably true to say that manual peptide and protein sequencing is currently practised worldwide in more laboratories than ever. The reasons are fairly obvious; the substantial initial capital outlay on automated equipment, the requirement for expert technical back-up and the high consumable and maintenance costs involved in its operation. Thus they are an economic proposition only when the number and type of samples to be sequenced justifies this large expenditure. Manual sequencing on the other hand uses no less sophisticated chemistry, but with a minimum of low-cost laboratory equipment such as may be found in any reasonably equipped biochemical laboratory.

In practice the only serious disadvantages of manual sequencing compared to automated sequencing are as follows.

(i) The repetitive yield is no more than 95%, compared with up to 98% for automatic sequencers. The practical consequence of this is that the length of N-terminal sequence that can be determined generally falls in the range of 10–20 residues depending upon the amount of starting material and the nature of the actual sequence being investigated (e.g. its hydrophobicity and solubility properties). The method is therefore best suited to peptides of this length rather than to whole proteins or long peptides. It is true that some authors have manually sequenced up to 50 residues into a peptide/protein, but these were exceptional achievements by workers with 'state-of-the-art' expertise, and not the normal expectation of the average newcomer to the field.

(ii) Manual methods are much more labour intensive. It is unlikely that an operator can manage more than 3–5 cycles, including identifications, in a working day, compared with 30 or more using the more recent automatic sequencers.

(iii) The most sensitive manual methods routinely require 0.5–5 nmol of starting material; cruder methods perhaps 5–50 nmol. However, automatic sequencers can operate successfully in the 1–20 pmol range, although routine sequencing is usually carried out in the 20–200 pmol range.

The major advantage of manual sequencing methods is that, in general, any laboratory boasting a decent fume-hood, a vacuum pump with multiple manifold, a refrigerator, one or two ovens or heating blocks, an adjustable micropipette (10–200 μl), a bench centrifuge and facilities for thin-layer chromatography (TLC), can realistically expect to begin to generate useful sequence data almost immediately. Manual sequencers have a second advantage over their automated counterparts in that it is perfectly feasible, in fact desirable for the sake of efficiency, to simultaneously sequence several peptides in a batchwise manner. Routinely in our laboratory, for example, two complete cycles a day, including identifications are carried out simultaneously on 16–32 samples.

The availability of only manual sequencing methods need not of itself seriously limit the scope of the investigations undertaken and the manual sequencer can apply himself to a wide range of problems.

(i) A unique N-terminal sequence (or at least N-terminal amino acid) is excellent confirmation that a protein/peptide preparation is homogeneous,. It is often the case that a protein that runs as a single band on polyacrylamide gel electrophoresis (PAGE), or even a peptide that elutes from a high-performance liquid chromatography (HPLC) column as a single sharp peak, is not homogeneous when subject to N-terminal analysis.

(ii) Comparison of a short N-terminal sequence with published sequences can aid positive identification of a protein.

(iii) Batches of peptides can be screened to determine which should be subjected to automatic sequencing.

(iv) N-terminal sequence determination may be used in conjunction with DNA sequence studies to establish initiation points and reading frames.

(v) The elucidation of short internal sequences of proteins may yield information which allows the construction of DNA probes.

(vi) The complete sequence can be obtained of small proteins (mol. wt 2000), for example, peptide hormones.

(vii) Finally, it is possible to elucidate the complete amino acid sequence of proteins of molecular weight up to 20 000 or 30 000 (or even larger) from the sequences of smaller overlapping peptides of a size suitable for manual sequencing, that is, up to 30 residues. The strategy used to derive suitable small peptides must be determined for each new protein (or protein family) but consists essentially of using a combination of specific chemical and/or enzymatic cleavages of the intact protein, or large fragments derived from it. The use of a number of these techniques is described below (Section 3.1.3) and in Chapter 2.

The potential usefulness of protein sequence analysis for a wide range of purposes, from evolutionary studies to protein engineering and the frontiers of molecular biology, coupled with the heavy commitment of resources required to run an automated sequencer, makes the acquisition of manual sequencing skills a valuable asset.

2. N-TERMINAL SEQUENCING METHODS: A PERSPECTIVE

All current N-terminal sequencing methods are based on the procedure devised in 1950 by Edman (1,2) and generally termed the 'Edman degradation reaction'. Essentially, as currently practised (3,4), this procedure (see *Figure 1*) consists of

(i) a *coupling* reaction carried out in alkaline conditions, and under nitrogen, in which the reagent phenylisothiocyanate (PITC) reacts with the N-terminal amino group to form an *N*-phenylthiocarbamoyl-(PTC) derivative of the peptide;

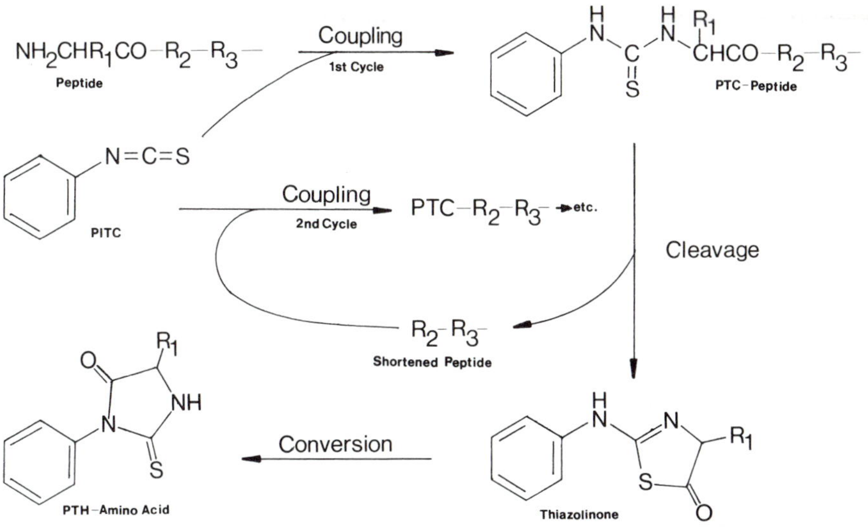

Figure 1. The Edman degradation reaction. PITC, phenylisothiocyanate; PTC, phenylthiocarbamoyl; PTH, phenylthiohydantoin.

(ii) a *washing* procedure to remove excess PITC and buffer;

(iii) a *cleavage* reaction in anhydrous acid conditions which causes this PTC-derivative to cyclize to yield a free thiazolinone corresponding to the original N-terminal amino acid and a peptide shortened by loss of its N-terminal residue;

(iv) the *extraction* of the thiazolinone into an hydrophobic solvent to separate it from the shortened peptide;

(v) the *conversion* of the unstable thiazolinone to the more stable phenylthiohydantoin (PTH) by treatment with dilute acid, before identification, usually by TLC or HPLC.

The shortened peptide can now be recycled through the Edman degradation procedure to yield the PTH-derivative of the next residue, and so on. This sequence of coupling, washing, cleavage, extraction and conversion is referred to as a 'cycle', and in practice, with practise, takes about 2 h to complete manually. Although a large number of important refinements to the details of the basic Edman degradation method have been introduced over the years, the most recent significant advances have largely been feats of technique and engineering, allowing for increased cycling efficiency and sensitivity of detection of the products.

The original Edman procedure depended upon back-hydrolysis of the PTH-derivative with barium hydroxide and identification of the parent amino acid by paper chromatography and required 1–2 mmol of starting material. Detection methods have improved dramatically over the following 20 years with the introduction of TLC, gas–liquid chromatography (GLC) and HPLC separations. Now the PTH-derivatives can be accurately determined directly in the 1–10 pmol range in 20–30 min. If, however, HPLC is not available, the current TLC methods are capable of resolving most PTH-amino acids satisfactorily in the low nmol range. Typically, ascending TLC is carried out on 15 cm silica gel plates pre-coated with fluorescent indicator. The chromatogram is generally developed in one dimension with two successive solvent systems in about 30 min. Eight to ten samples and standards may be run together on one plate. After drying the chromatogram is photographed under UV light (254 nm) although the plates should not be exposed to prolonged illumination with UV since the spots do fade. Unfortunately, the Edman procedure generates appreciable amounts of a number of UV-absorbing by-products of many amino acids as well as large amounts of phenyl- and diphenylurea, all of which may obscure the results and make positive identifications more difficult. Additional information can be obtained by spraying the sheets with freshly prepared 0.1% ninhydrin, 5% collidine in ethanol and heating at 110°C for 5 min when many of the PTH-derivatives develop specific colours which aid their identification. Unfortunately, all information must be recorded almost immediately since the plates quickly turn completely red and are unreadable within 30 min.

An important modification of the basic Edman technique arose following the development of the protein label 5-dimethyl-amino-naphthalene-1-sulphonyl chloride (dansyl chloride) (5). This reagent reacts with free amino groups of peptides and proteins to form acid stable dansyl (DNS) derivatives. Total hydrolysis of the dansylated peptide yields a mixture of free amino acids together with the DNS-derivative of the original N-terminal amino acid. Since this DNS-amino acid is very strongly fluorescent in UV light it can be easily identified, and with great sensitivity, after separation by two-dimensional TLC. Hartley *et al.* combined the Edman and dansylation procedures into the dansyl–Edman Method (6), in which the N-terminus of the peptide or protein is determined by dansylation of a small sample, whilst the rest is subjected to the usual Edman procedure of coupling, washing, cleavage and extraction. The extracted thiazolinone derivative is usually discarded. A few percent of the shortened peptide is then taken and its newly exposed N-terminus determined by dansylation whilst the rest is subjected to another Edman cycle.

The loss of a small percentage of the peptide at each cycle for the dansylation reaction is easily compensated for by the increased (almost two orders of magnitude) sensitivity of TLC methods for identifying DNS-amino acids over similar methods for identifying PTH-amino acids. The method does, however, have a number of drawbacks. The most serious is that during the total hydrolysis step DNS-Asn and DNS-Gln are completely deamidated to yield DNS-Asp and DNS-Glu respectively. Definitive assignments cannot then be given. There are also problems with DNS-Trp, -Pro, -Ser, -Thr; the first is completely destroyed on acid hydrolysis and the yields of the rest, particularly DNS-Pro (at ~30%), are significantly reduced unless short hydrolysis times are used. Also, if a peptide has two successive hydrophobic N-terminal amino acids, or the second residue is a proline, there is a tendency to get high yields of the DNS-dipeptide on hydrolysis causing problems with identification. The use of short hydrolysis times to improve the yield of unstable DNS-derivatives may in turn aggravate this problem.

In order to overcome the disadvantages of the dansyl–Edman approach, a number of substituted PITCs have been investigated from time to time. In recent years Chang and co-workers have introduced a coloured isothiocyanate derivative [4,*N*;*N*-dimethyl-aminoazobenzene-4′-isothiocyanate, dabsyl-isothiocyanate or DABITC (*Figure 2*)]. This reagent can partly replace PITC in the coupling reaction leading eventually to the release of intensely red-coloured thiohydantoin derivatives, generally termed DABTH-amino

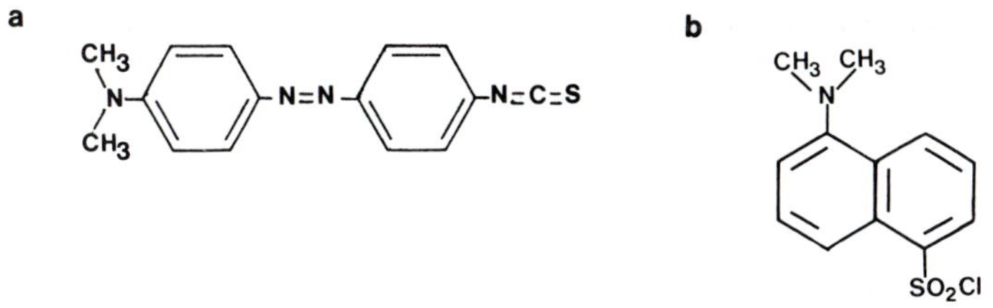

a

CH_3 N CH_3 — N=N — N=C=S

b

CH_3 CH_3 N — SO_2Cl

Figure 2. (**a**) 4,*N*,*N*,-dimethylaminoazobenzene-4′-isothiocyanate (DABITC: dabsyl isothiocyanate). (**b**) 1-Dimethylaminonaphthalene-5-sulphonyl chloride (DNS-Cl or dansyl chloride).

acids, which are readily separated and visualized by TLC (7,8). The level of sensitivity is similar to that for DNS-amino acids. The only drawback with this modified isothiocyanate is that under the conditions of temperature and reaction time normally used with the Edman degradation, it couples with an efficiency of only about 60% compared with essentially 100% for PITC. Thus, if DABITC alone were used in the coupling reaction, then at the beginning of the second cycle there would be a 40:60 mixture of the original peptide and the peptide shortened by one residue and therefore two DABTH-derivatives would be found after the second cycle. Thereafter the problem escalates rapidly and the definitive identification of the sequence becomes impossible after only a few cycles. The simple answer to this problem was to include a second coupling step using PITC following that with DABITC to ensure 100% derivatization and no carry-over (9). The colourless PTH-derivatives and other UV-absorbing by-products do not interfere with the subsequent identification of the coloured DABTH-derivatives. This most elegant DABITC/PITC double-coupling method is arguably the most satisfactory and is currently the method of choice for most laboratories involved in the manual sequencing of peptides and proteins. Although the statement clearly reflects the author's own experiences and prejudices, it is the method to be most strongly recommended to newcomers to the field.

Although originally developed as a liquid-phase sequencing method, and most often used as such, the DABITC/PITC double-coupling method may also be used in conjunction with manual or automated solid-phase sequencing methods (10,11 and see Chapter 3). Such an approach can be useful for sequencing hydrophobic peptides which might otherwise be 'washed-out' during the organic extractions of the liquid-phase method. Consequently, protocols for protein and peptide attachment to inert supports and for subsequent sequencing are included below. All peptides having a lysine, a free C-terminal carboxyl group or a C-terminal homoserine residue can be attached according to the methods described, generally with a yield of 30–80%.

3. DABITC/PITC DOUBLE-COUPLING METHOD

3.1 Materials and reagents

This section contains details of all chemicals, reagents and equipment required for routine manual sequencing in the 1–10 nmol range. Where more material is available (10–100 nmol), less pure chemicals may be adequate but the techniques can only be applied successfully in the range 250 pmol–1 nmol provided that very pure reagents are used and strict attention to detail is maintained. Where the recommended grades are not readily available or too expensive for routine use, re-distillation or re-crystallization is usually necessary.

3.1.1 *Chemicals and solvents*

(i) *Pyridine, PITC and trifluoroacetic acid (TFA).* These reagents should all be of sequencer grade. Aqueous solutions should be made up using the best quality double-distilled water (or better). Once ampoules of PITC have been opened, the contents should be transferred to re-sealable tubes and stored at −20°C under nitrogen.

(ii) *DABITC.* This can be used directly but cleaner chromatograms may be obtained if this is re-crystallized from boiling acetone (1 g in 70 ml). DABITC is dissolved as required in anhydrous pyridine at 2.8 mg/ml and used directly. Alternatively, it may be dissolved in bulk at the same concentration in anhydrous acetone, and 0.25 or 0.5 ml aliquots dispensed into small tubes. These are dried down under vacuum and stored at −20°C in a desiccator. One or more tubes may then be reconstituted with pyridine as required.

(iii) *Heptane, ethyl acetate, n-butyl acetate, toluene and acetic acid.* These should all be of analytical reagent grade or better.

These reagents can be obtained from one or more of Rathburn, Fluka, BDH or Pierce.

3.1.2 *Equipment*

(i) *Reaction tubes.* Tubes used for the sequencing reactions should be of thick-walled borosilicate glass and fitted with ground-glass stoppers. The dimensions are to some extent determined by the rotor available for the bench centrifuge used during the washing and extraction stages, but they should be approximately 50−60 mm long with an internal diameter of not more than 8 mm. Ideally, the bottoms of the tubes should be slightly tapered. The stoppers should extend at least 1.0 cm above the ground glass joint for ease of handling. [In our laboratory sequencing tubes are constructed from Quickfit sockets (SRB7) and long-tipped cones (CBB7) which are widely available.] Tubes for solid-phase sequencing should be somewhat larger (70 × 10 mm) and require an additional sintered glass stopper for use whenever the tubes are placed under vacuum. These restrict the vacuum and reduce potential losses of support bead due to 'bumping'.

(*Note*, limited success has been reported by colleagues using 'Eppendorf' snap-top 1.5 ml centrifuge tubes. These do have the disadvantage of not being completely transparent, and are prone to 'bumping' during evacuation.)

The dimensions of the conversion tubes are not critical. 60 × 10 mm screw top glass test-tubes are ideal but a variety of tubes can be used here provided that they are sealable, heat-resistant and do not induce 'bumping' when placed under vacuum.

(ii) *Temperature maintenance.* All protocols require elevated temperatures during the coupling and cleavage reactions. This can best be achieved via a series of thermostatically heated blocks which provide a snug fit for the sequencing tubes. Excellent results can also be achieved using water baths, or efficient fan ovens. Temperatures for coupling, cleavage and conversion reactions do not seem to be critical with 50−55°C being used for coupling and cleavage and 55–80°C for conversion in different laboratories. The protocols described below recommend 52 and 80°C respectively.

(iii) *Vacuum pumps.* A good vacuum system is *essential.* Ideally two oil pumps should be used. The first for the removal of volatile bases and solvents after coupling and extractions, and the second reserved for removal of TFA after cleavage and conversion. The vacuum pumps should be protected by a cryostat and the TFA pump should also be protected by a KOH trap placed between it and the cryostat. Each vacuum pump can be connected to a manifold so that several desiccators can be accommodated simultaneously. If absolutely necessary both manifolds may be connected to the same pump, however, extreme care should then be taken to avoid acid fumes entering desiccators containing base, and vice versa. It is clear that salt formation occurring under these conditions, or introduced with the sample, is a major contributor to poor sequencing efficiency and low repetitive yields, resulting in subsequent failure to sequence to the end of even short peptides. Salt accumulation during sequencing also leads to progressive difficulties in running clean chromatograms and particularly in resolving histidine and arginine.

(iv) *TLC sheets and chromatography equipment.* Polyamide thin-layer sheets (150 × 150 mm) can be purchased from BDH or Schleicher and Schüll and are carefully cut into 30 × 30 mm sheets for chromatography of DABTH-derivatives. Suitable small solvent chambers for the 30 × 30 mm sheets can be constructed from cut-off 50 ml glass beakers sealed with glass plates or inverted larger beakers. Simple stainless steel racks capable of holding 10–20 of the 30 × 30 mm sheets are convenient for holding the sheets during drying between solvents. They can also be used for simultaneous chromatography of a number of samples (however, see Section 3.4.3).

3.1.3 *Sample preparation*

Notwithstanding all that has previously been said about the sensitivity of the various sequencing methods for individual peptides, workers inexperienced in the techniques of peptide purification and sequencing would be ill-advised to embark upon the determination of a complete amino acid sequence on an unknown protein, without access to at least 0.5–1 μmol of starting material. This obviously should be as homogeneous as possible by at least two independent criteria although a few percent protein impurities are not likely to cause problems. The fact that a protein elutes as a sharp peak from a size exclusion column and runs as a single band on sodium dodecyl sulphate (SDS)–PAGE is not of itself adequate proof of purity and a better test is the demonstration of a unique N-terminal amino acid, or preferably a short N-terminal sequence.

The basic manual sequencing strategy is to establish as much of the N-terminal sequence as possible on a few nmol of the whole protein and then to selectively cleave a sample of the protein into peptide fragments each of such a size that, once purified, they can be effectively sequenced to the end. Alignment of the individual peptide sequences is only possible after sequencing a second set of peptides produced using a different cleavage method, such that the two sets of sequences will overlap. With luck, but rarely, two digests could be sufficient. Much more often additional digests, including sub-digests (e.g. peptides produced via one enzyme are re-digested with a second enzyme) will be necessary and the actual choice of cleavage method to be used in any given case should only be made after detailed consideration, particularly when weeks of work may have gone into producing a few precious milligrams of protein! Details of recommended methods of enzymic and chemical cleavage are given in Chapter 2.

Before attempting to fragment the protein it is essential to establish its molecular weight and obtain an accurate amino acid composition in order to allow the correct choice of cleavage strategy. If cysteine is present then the protein should usually be reduced and the sulphydryls modified prior to digestion (see Section 6.1). It is important to appreciate that the occurrence of particular residues in the amino acid composition does not guarantee the presence of useful or susceptible cleavage sites. When the sequence is complete, the total number of amino acid residues found should correspond to the molecular weight and the amino acid composition of the protein. Similarly, N-terminal sequencing of individual peptides is greatly facilitated if the amino acid composition of the peptide is known. Where an amino acid analyser is not available, semi-quantitative amino acid compositions of short peptides can be determined by TLC following total acid hydrolysis and dansylation (see Sections 4 and 6.3).

Classically, the cleavages performed were chosen so as to yield the smallest number of peptides of the largest size suitable for sequencing. One major reason for this was that the methods for purifying peptides (preparatively) were often laborious and constituted the principal rate-limiting step in the sequencing methodology. Frequently large and/or hydrophobic peptides would be quite intractable or be lost altogether. In contrast, the currently accepted strategy, especially for manual sequencers capable of sequencing dozens of samples simultaneously, is to produce a relatively large number of small (10–20 residue) peptides and to use the high resolving power, sensitivity and rapidity of HPLC for their purification (see ref. 12 for further discussion). Since mixtures of more than 20 peptides are unlikely to be completely resolved in a *single* HPLC run, a preliminary resolution by size exclusive column chromatography is sometimes advisable. We routinely use a 1 × 200 cm column of Bio-Gel P6 or P10 (Bio-Rad) in 0.05 M ammonium bicarbonate (pH 8.1) or 70% formic acid to achieve a preliminary separation of complex mixtures (13). Peak fractions are pooled to give 8–10 samples which are subsequently subjected to reverse-phase HPLC. Generally excellent results are obtained with 25 × 0.5 cm C18 Vydac, μ-Bondapak (Technicol Ltd, Macclesfield, Cheshire) or Micro-Pac MCH-10 resins (Varian Assoc. Inc.), using linear gradients of 0–70% acetonitrile in 0.1% TFA and a separation time of 60–120 min depending on the complexity of the mixtures. Peptides are detected by monitoring at 214 nm. This combination of conventional and HPLC resins has the additional advantage that volatile solvents are employed so that the final peptides can be lyophilized without desalting and then dissolved directly in 50% pyridine for sequencing.

3.2 **Liquid-phase manual sequencing**

3.2.1 *Protocol*

(i) *Coupling.*

(a) Dissolve the sample (5–10 nmol) in 80 μl of 50% pyridine; add 40 μl (400 nmol) of DABITC (2.8 mg/ml in pyridine); purge with N_2, seal the tubes and incubate at 52°C for 50 min.

(b) Add 10 μl of PITC; purge with N_2; seal the tubes and mix thoroughly (vortex); incubate at 52°C for 20 min.

(ii) *Washing.*

(a) Add 500 μl of heptane:ethyl acetate (2:1), vortex and separate the phases by centrifuging briefly (1 min).
(b) Aspirate off the upper phase and discard (see Section 3.2.2).
(c) Re-extract the lower (aqueous) phase with heptane:ethyl acetate twice as above, discarding the upper phase on each occasion.
(d) Dry the aqueous phase under vacuum. (It is most important that the samples are *completely* dry before proceeding to the cleavage step.)

(iii) *Cleavage.* Add 50 μl of anhydrous TFA; purge with N_2; seal the tubes and incubate for 15 min at 52°C. Remove the TFA by drying under vacuum.

(iv) *Extraction.*

(a) Add 200 μl of *n*-butyl acetate and 50 μl of water; vortex well and centrifuge briefly.
(b) Remove the upper (organic) phase into a conversion tube.
(c) Dry the extract and the aqueous phase under vacuum.
(d) The dried 'aqueous phase' may now be subjected to a further cycle; the thiazolinone, dried down from the organic phase, is converted to the hydantoin.

(v) *Conversion.*

(a) Add 50 μl of 50% TFA to the conversion tube, seal and incubate at 80°C for 10 min.
(b) Dry under vacuum.
(c) Re-dissolve in 2–5 μl of ethanol and take a suitable aliquot for polyamide TLC.

(vi) *Modifications for N-terminal analysis only.* If the N-terminal amino acid only is to be determined then the second coupling with PITC is omitted.

3.2.2 *Notes on protocol*

(i) Salt contamination reduces repetitive yield and increases the cycle time by increasing the time required for drying. Accordingly, methods of protein and peptide purification should be chosen which will minimize the salt content, for example by the use of volatile buffers. Ammonium sulphate precipitation should be avoided wherever possible and the use of SDS is inadvisable. Where salts are present at least two independent de-salting steps are recommended prior to sequencing. Dialysis alone is rarely sufficient. Peptides purified by reverse-phase HPLC using gradients of organic solvents such as acetonitrile and then lyophilized yield excellent results. Progressive salt accumulation may occur during sequencing unless care is taken to ensure that after the washing stage, the lower aqueous phase is *completely dry* before addition of TFA for the cleavage reaction. Salt accumulation can also occur if desiccators containing samples in TFA and pyridine are attached to the same vacuum line (see Section 3.1.2).

(ii) During the washing stage, great care should be taken not to disturb the interface between the organic and aqueous layers after centrifugation. Some peptides become insoluble during this washing stage and may be found as a precipitate at this interface. For this reason always leave 1−2 mm of the organic phase after each extraction. Should the organic phase retain an appreciable yellow colour (due to DABITC) after the third extraction, then an additional extraction step will be necessary. With practice, three extractions should be sufficient.

(iii) It is recommended that the cleavage step be repeated two or even three times if a peptide bond involving particularly hydrophobic residues is suspected (e.g. involving Pro, Val and Ile). This is particularly important if the bond is near the beginning of a long peptide.

(iv) Fast drying times *are* important and an efficient vacuum system must be maintained.

(v) If it is neccessary to interrupt a cycle, this is best done after removal of TFA in the cleavage stage. The cycle should be completed the next day. Samples are otherwise best left in 50% pyridine ready to start the coupling reaction, rather than being stored dry.

(vi) Removal of heptane:ethyl acetate is best achieved using a fine Pasteur pipette connected to a water vacuum pump via a trap to collect the waste solvent.

3.3 **Solid-phase manual sequencing** (see also Chapter 3)

Three main attachment procedures are commonly used. The first is suitable for proteins and for peptides having a lysine residue, and involves attachment via the ϵ-amino group to *p*-phenylenediiosothiocyanate (DITC)-activated aminopropyl-glass (APG) beads. Yields are generally good and may exceed 80%. Since the N-terminal amino group is also mostly attached, the first residue may not be detected in the sequencing procedure since it remains bound to the support after treatment with TFA. Intrachain lysine residues similarly remain attached to the support and result in gaps in the sequence if attachment is essentially quantitative. Obviously a C-terminal lysine also is undetectable. This attachment method will tolerate reasonable amounts of salt (*not* NH_4^+) and SDS but later sequencing yields may be adversely affected.

Secondly, small peptides with a free C-terminal carboxyl group may be attached to APG beads after carboxyl activation with 1-ethyl-3(3-dimethylaminopropyl)-carbodiimide hydrochloride (EDC). The C-terminal residue should *not* be lysine, proline or glutamine. No salt, SDS or acid is tolerated and yields are highly variable (20−70%). The method often fails or gives very low yields with large peptides or proteins, possibly due to solubility problems. Attachment of side chain carboxyl groups (even those of N-terminal aspartate or glutamate) is not complete, and sufficient remain free to allow subsequent identification during sequencing.

The third method involves attachment to APG beads of peptides having a C-terminal lactone derived from methionine or tryptophan residues following chemical cleavage. This method (see also Section 3.2.2, Chapter 3) is very efficient.

3.3.1 *Attachment to supports* (10)

(i) *Preparation of DITC-activated glass beads.*

(a) Slowly add 2 g of APG beads (Fluka; 170 Å; 200−400 mesh) to a solution of

1 g of DITC in 13 ml of dimethylformamide (DMF); stir gently for 2–3 h at room temperature.

(b) Collect the derivatized beads on sintered glass filters and wash with 20 ml of DMF followed by 20 ml of methanol.

(c) Dry under vacuum at room temperature and store in a refrigerator under nitrogen.

(ii) *Attachment of peptides to DITC–glass.*

(a) Dry 5–10 nmol peptide in a solid-phase sequencing tube under vacuum. Re-dissolve in 400 µl of 0.4 M $NaHCO_3$, 10% isopropanol (up to 1% SDS may be included to aid solubility). Add 10 mg of DITC–glass beads (freshly prepared if possible); flush with N_2; de-gas briefly, cap the tube and stir gently at 50°C for 1 h (see Section 3.3.3).

(b) Add 20 µl of ethanolamine; flush with N_2, and incubate at 50°C for 15 min.

(c) Recover the beads by centrifugation; wash them twice with water and twice with methanol. The attached peptides may now be sequenced as below, or stored after drying under vacuum.

(iii) *Attachment of C-terminal carboxyl groups to APG.*

(a) Dry 5–10 nmol of peptide in a solid-phase sequencing tube under vacuum, add 100 µl of anhydrous TFA, flush with N_2 and incubate for 15 min at room temperature.

(b) Remove the TFA under vacuum. Add 2 mg of EDC in 200 µl of DMF followed by 10 mg of APG, flush with N_2, de-gas *in vacuo* (water pump; do not remove solvent) and incubate with gentle stirring for 1 h at 40°C (Section 3.3.3).

(c) Recover the beads by gentle centrifugation and wash twice with 1 ml of water and twice with 1 ml of methanol.

(d) Add 100 µl of 20% PITC in DMF and 200 µl of 25% pyridine in DMF, flush with N_2, de-gas and incubate at 50°C for 20 min.

(e) Recover the beads by gentle centrifugation, wash them twice with DMF and twice with methanol. The attached peptides can now be sequenced as below or the beads may be stored after drying under vacuum.

(iv) *Attachment of C-terminal homoserine lactone to APG.*

(a) Dry 5–10 nmol of peptide in a solid-phase sequencing tube under vacuum. Add 300 µl of TFA, flush with N_2 and incubate for 1 h at room temperature. Dry under vacuum. (This quantitatively converts all homoserine to the lactone.)

(b) Re-dissolve in 300 µl of DMF, add 10 mg of APG and 50 µl of triethylamine, flush with N_2, de-gas and incubate with gentle stirring for 2 h at 45°C (see Section 3.3.3).

(c) Recover the beads by centrifugation, wash them twice with DMF and twice with methanol. The attached peptides may now be sequenced as below or may be stored after drying under vacuum.

3.3.2 *Protocol*

(i) *Coupling.*

(a) Add 200 µl of 50% pyridine to the solid-phase sequencing tube containing the peptide attached to the glass beads; add 100 μl of DABITC (2.8 mg/ml pyridine) and purge with N_2 for 10 sec. Seal the tubes and incubate them at 50°C for 50 min (see Section 3.3.3).

(b) Add 20 μl of PITC, flush with N_2 and incubate at 50°C for 30 min.

(ii) *Washing.* Centrifuge and aspirate off the supernatant. Wash the beads twice with 500 μl of pyridine followed by twice with 500 μl of methanol. Insert sintered stoppers and dry under vacuum.

(iii) *Cleavage.* Add 200 μl of TFA, flush with N_2, seal the tubes and incubate them at 50°C for 10 min. Replace the stoppers with sintered stoppers and dry under vacuum.

(iv) *Extraction.* Add 400 μl of methanol, centrifuge and transfer the supernatant to the conversion tube. Repeat the extraction. Dry the combined extracts and glass beads under vacuum. The dry beads may now be re-cycled.

(v) *Conversion.* Add 50 μl of 50% TFA to the dried extract, cap the tubes and incubate at 80°C for 10 min. Dry under vacuum. Re-dissolve in 2–5 μl of ethanol and take a suitable aliquot for analysis by polyamide TLC.

3.3.3 *Notes on protocols*

(i) Manual solid-phase sequencing has not achieved the popularity of the liquid-phase method, although potentially it should be possible to sequence somewhat longer peptides this way and there should be no losses of small hydrophobic peptides as sometimes occurs during the washing stages of the liquid-phase method. However, the solid-phase approach suffers the disadvantage of material loss during the various modification and coupling procedures. Since these losses are quite unpredictable, it has usually been regarded as unacceptable to risk losing more than 50% at the attachment stage. However, if difficulties are encountered in sequencing hydrophobic peptides by the liquid-phase method, or if large amounts of relatively long peptides (30–40 residues) are available, the increased repetitive yield may justify some experimentation to determine the ideal attachment conditions for individual peptides. Otherwise the liquid phase system is recommended.

(ii) Potentially the greatest problem with this sequencing method is loss of glass-attached peptide due to bumping during the drying phases. This is best avoided by the use of special sintered glass stoppers and by evacuating desiccators for 2–3 min using a water vacuum pump before attaching to the oil vacuum pump.

(iii) The chemistry proceeds more efficiently if the contents of the sequencing tubes are stirred gently but constantly. This is best achieved by incorporating a stirring bar in each sequencing tube and placing a magnetic stirrer below the heating block.

3.4 Identification of DABTH-amino acid derivatives

Identification of individual DABTH-amino acids may be achieved by reverse-phase HPLC in about 30 min on a 250 × 4.5 mm C8 column [5 μm MOS-Hypersil (Shandon)] using isocratic elution with 12 mM sodium acetate pH 5.0 acetonitrile (1:1 v/v) containing 0.5% 1,2-dichloroethane at a flow-rate of 1.2 ml/min and at 45°C (14). The standard method has the disadvantage that DABTH-arginine is not eluted from the column, although it can be determined in a separate run using 12 mM sodium acetate pH 5.0:acetonitrile (1:4 v/v) containing 0.5% 1,2-dichloroethane. Alternatively, two-dimensional TLC on 30 × 30 mm polyamide sheets can be used (9). This very simple, efficient and inexpensive technique ideally complements the manual DABITC/PITC double-coupling method and is described in detail below.

(*Note*: Often 25 × 25 mm polyamide sheets are used but, in our experience, the convenience and resolution of the slightly larger sheets far outweighs the additional cost. However, if one is working at the lowest limits of sensitivity of this method (10 pmol), the *smaller* sheets will give more discrete spots and so aid identification.

3.4.1 *Chromatography markers*

Chang (9) recommends the use of a mixture of diethylamine and ethanolamine reacted with DABITC as chromatography markers against which the mobilities of the DABTH-amino acids are compared. However, once experience has been gained in recognizing the characteristic spot patterns produced by the amino acid derivatives, the ethanolamine standard can easily be dispensed with. This can be an advantage since the DABTH-derivative of carboxymethylated cysteine (CM) runs in the same position. The chromatography markers are prepared as follows.

(i) Pipette 50 μl of pyridine into an Edman tube and add 30 μl of diethylamine (and 30 μl of ethanolamine if required) and 250 μl of DABITC solution (2.8 mg/ml in pyridine).
(ii) Stopper the tube and incubate at 52°C for 1 h.
(iii) Dry under vacuum.
(iv) Dissolve in 1 ml of ethanol and store in a screw-top test-tube until required.

The DABITC-reacted diethylamine/ethanolamine is very stable (at least several months) and if it dries out, can easily be reconstituted by adding ethanol. One small spot (perhaps 0.5 μl) of the markers is all that is required for each chromatogram. The spots should be easily discernible but nothing is to be gained from overloading, indeed sample spots may be obscured.

3.4.2 *Preparation of standard DABTH-amino acids (15)*

(i) Dissolve 1 mg (~5 μmol) of amino acid in 100 μl of TEM buffer (50 ml of water, 50 ml of acetone, 0.5 ml of triethylamine, 5 ml of 0.2 M acetic acid, pH 10.65) and add 50 μl (200 nmol) of DABITC (1.13 mg/ml in acetone).
(ii) Seal the tubes and incubate them at 52°C for 1 h.
(iii) Dry *in vacuo*.
(iv) Re-dissolve in 50 μl of 50% TFA and incubate at 52°C for 45 min (or 80°C for 10 min) and again dry *in vacuo*.
(v) Dissolve in ethanol at a concentration of 5–10 pmol/μl.

3.4.3 *Chromatography*

Chromatography is carried out on 30 × 30 mm polyamide sheets. These may be labelled lightly in pencil in their top right-hand corners to identify peptide and cycle number.

Care should be taken not to break the layer.

(i) Spot the chromatography marker and samples in the bottom left-hand corner about 4 mm from either edge. The amount of sample to be run will depend upon the concentration of the original peptide and will increase with increasing number of degradative cycles performed due to the cumulative effect of extraction losses.

(ii) Where more than one aliquot has to be applied allow the first to dry properly before adding the second and take care to keep the spot size down to about 1 mm diameter or less. This is greatly facilitated by spotting under a current of warm air from a hair dryer. After conversion, the dried DABTH-derivatives are red to purple in colour and when dissolved in ethanol the solution is bright yellow and dries *brown* on the TLC sheet. (Dark purple spots at this stage will invariably result in unsatisfactory chromatographic resolution. If this happens, remove the ethanol under vacuum, add 50 μl of water and 200 μl of butyl acetate, vortex well and centrifuge briefly. Remove the upper phase to a fresh conversion tube and convert as described earlier. Dry well under vacuum and re-dissolve in 2–3 μl of ethanol for chromatography.) The intensities of these colours are, of course, proportional to concentration and experience will soon allow determination of the correct volume of sample to be spotted but, as a guide, 50–100 pmol of DABTH-derivative should give unambiguous results.

(iii) The chromatogram is developed in two dimensions in covered tanks. Run in the first dimension with the sample in the bottom right-hand corner; the solvent is glacial acetic acid:water; 1:2 (v/v). Allow the solvent to run to the top edge of the sheet (2–3 min).

(iv) As soon as the solvent reaches the top of the sheet, remove the sheet from the solvent and dry thoroughly under a current of warm air, (i.e. until there is no smell of acetic acid).

(v) Turn the sheet through 90° and run in the second dimension in toluene:*n*-hexane: acetic acid; 2:1:1 (by vol.) until the solvent front is 2–3 mm from the top of the sheet (1–2 min).

(vi) Remove the sheet and dry under a current of warm air.

The same solvents may be used for several consecutive sheets but must be changed for each set of identifications. The second solvent is particularly volatile and prone to 'ageing' with consequent changes in mobilities of hydrophobic spots which tend increasingly to run with the solvent front.

It is possible to construct simple racks to allow several sheets to be chromatographed together in the same tank. In our experience these do not greatly increase efficiency and may result in solvent creeping up between rack and sheet and ruining the chromatogram.

3.4.4 *Identification*

DABTH-derivatives are visualized by exposing the chromatograms to concentrated HCl fumes when they become visible as red spots. Stainless steel racks containing several sheets may be placed in a desiccator over concentrated HCl for a few seconds, or individual sheets may be held in a stream of acid fumes produced by directing a gentle stream of air into the neck of a Winchester containing about 100–200 ml of concentrated HCl. Care should be taken not to get acid directly on the sheets as this will dissolve them.

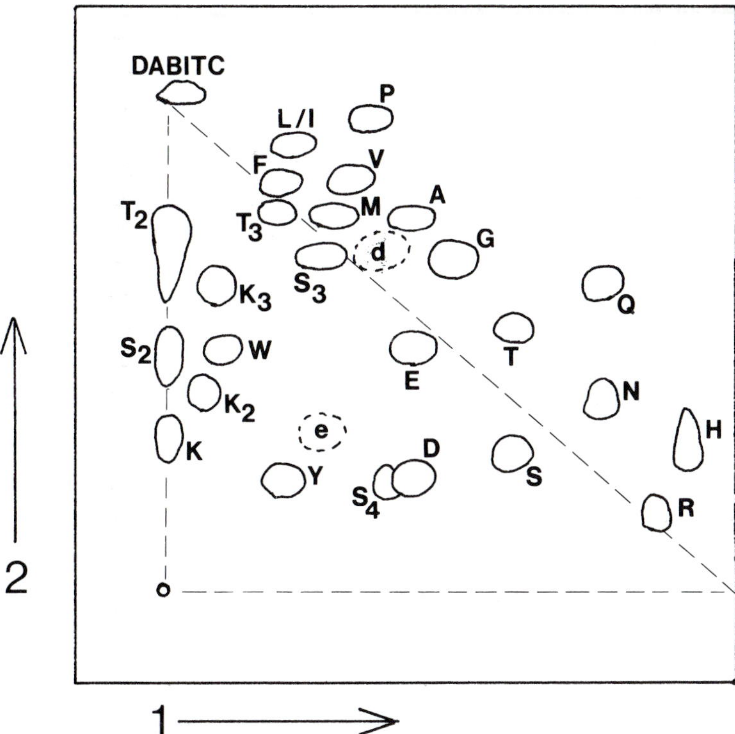

Figure 3. Schematic representation of 30 × 30 mm two-dimensional thin-layer chromatograph of major DABTH-amino acid derivatives to show their relative mobilities. Chromatography conditions are given in the text. The primary amino acid derivatives are identified by means of the single letter amino acid code and derivatives. T_2 and T_3, and S_2, S_3 and S_4 are common breakdown products of threonine and serine respectively, resulting from the use of anhydrous TFA in the cleavage reaction during sequencing. They do not occur during the preparation of DABTH-standards as described in Section 3.4.2. T_2 and S_2 are often the most prevalent forms found during sequencing. T_2 is blue/purple in colour; all other derivatives shown are red. K, K_2 and K_3 are all lysine derivatives as described in the text. K represents α-DABTH-ϵ-DABTC-lysine, normally found in the standard. K_2 and K_3 are mixed DABITC/PITC derivatives resulting from the double-coupling method used and are only found, together with K, during sequencing. 'd' and 'e' show the positions of the diethylamine- and ethanolamine-derived reference markers, and are purple. CM-cysteine runs underneath the ethanolamine standard and is not shown. The DABITC spot results from incomplete extraction during the washing phase and may be absent or present to a greater or lesser extent.

If the standard DABTH-derivatives are prepared from pure amino acids (Section 3.4.2) and subjected to chromatography as above then single spots are invariably found for most amino acids as shown in *Figure 3*. [The relative mobilities are slightly different from those published by Chang (9,15) probably due to the larger sheets used here.] Ideally the same result might be expected for each amino acid derivative obtained by the manual sequencing method. In practice, however, single spots are rarely achieved unless extremely strict attention is paid to reagent purity and to every detail of the sequence methodology. Even then, some experimentation will probably be required to determine the optimal conditions for individual laboratories. Instead of a single product, each amino acid usually gives rise to one or more additional spots due to components arising from incomplete conversion or partial acid destruction. The incidence and intensity of these secondary products generally increases with the number of degradative cycles performed. Some of these additional spots are blue or grey instead of red. The occurrence of secondary products does mean that the sensitivity of the method is reduced, perhaps by a factor of 2–3. However, since the secondary spots are themselves specific to individual amino acids they are in fact often a valuable aid in the identification of the amino acid concerned. For example the DABTH-derivatives of Leu/Ile, Met, Val, Phe and, to a lesser extent, Pro all run to positions within a couple of millimetres of each other on the chromatogram, just above the diethylamine marker. Considerable experience is required to identify unambiguously any one of these. In practice, however, instead of a single spot, one may have 2–4 spots of various colours, forming a characteristic 'fingerprint' for each residue, thereby greatly aiding unambiguous identification. Some of the secondary spots, particularly those associated with threonine, serine, lysine and the amides can be ascribed to identifiable breakdown or by-products (9,15); the rest remain unidentified.

The plate (*Figure 4*) is of sheets which have been selected to show most of the diagnostic spots that may be found for each residue. Note, however, that the relative intensities of the spots are subject to variation and all will not appear every time. The notes which follow are intended as an additional aid to identification of specific amino acids and should be read in conjunction with the plate and *Figure 3*. The photographs are of actual chromatograms obtained during peptide sequencing (i.e. they are not standards) and therefore spots due to carry-over also occur in some cases.

(i) *Proline* characteristically shows a double, rather diffuse, red spot running just behind the second solvent front, and generally a blue spot in the bottom right quarter of the chromatogram in the position shown for the DABTH-serine standard in *Figure 3*. An additional weak pink spot may sometimes be seen below and to the left of the diethylamine standard.

(ii) *Leucine/isoleucine* derivatives are not resolved from each other under these conditions. They can be positively identified by HPLC (14) or if sufficient material is available, by one-dimensional TLC on 10 × 10 cm silica gel sheets using formic acid:water:ethanol; 1:10:9 (by vol.) as solvent (16). This method is not very sensitive and often does not yield unambiguous results. In practice, it may not be necessary to resolve leucine/isoleucine in this way unless the amino acid composition of the peptide shows that both are present. Often, as a result of using a variety of cleavage methods to generate overlapping peptides, some peptides are obtained which will contain only Leu *or* Ile residues. (Also in this respect it is worth remembering that pepsin and chymotrypsin normally cleave on the C-terminal side of leucine but not isoleucine.) DABTH-Leu/Ile characteristically appear as two red spots above the diethylamine standard, with the major spot not overlapping the standard and the secondary spot essentially directly above the standard. In addition, a pink and a blue spot are often found in the lower half of the chromatogram, directly below the red primary spot, the three forming a vertical series with the blue spot at the bottom. These greatly aid the distinction between Leu or Ile, phenylalanine and valine.

(iii) *Phenylalanine* gives a double red spot in a similar place to those of Leu/Ile, although the principal spot is slightly lower, and both are a little more to the left of the diethylamine standard than is the case with Leu/Ile. Identification is

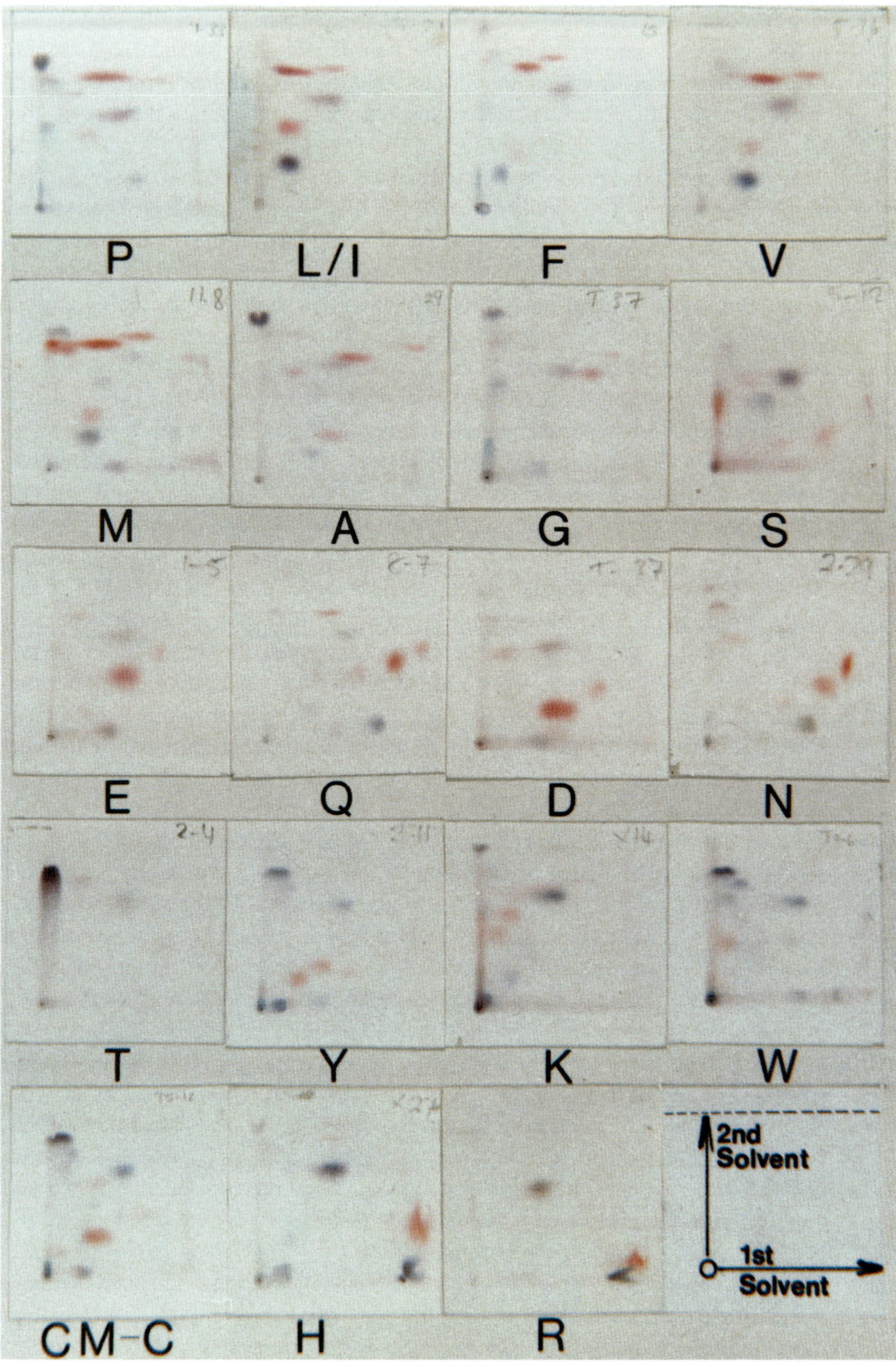

Figure 4. Chromatography of the DABTH-derivatives of amino acids, obtained by means of the DABITC/PITC double-coupling sequencing strategy. Solvent 1: glacial acetic acid:water (1:2 v/v); solvent 2: toluene: *n*-hexane:glacial acetic acid (2:1:1 by vol.). DABTH-derivatives are identified by the single letter amino acid code.

greatly aided by the pink and blue spots often found in the bottom left-hand corner of the chromatogram. Although generally similar in position and colour to the Leu/Ile derivatives described above, note that they are slightly closer together, lower down, and, in particular, that the blue spot is very much closer to the left-hand side, having hardly moved in the first solvent.

(iv) *Valine* also generally appears as two red spots but the major product is now further to the right so that it partially overlaps the diethylamine marker. Again there may be pink and blue spots in the bottom left-hand quarter of the chromatogram but note this time that the blue spot has moved significantly in the first dimension and the pink spot has moved slightly further so that these two form a *linear* but not *vertical* series with the principal spot as is seen with Leu/Ile.

(v) *Methionine* identification often causes most difficulty within this series of hydrophobic residues. Again it gives a double red spot with the major product overlapping the diethylamine standard, rather like valine but much nearer the standard. In addition, when present, the lower pink and blue spots are closer together and appear rather more staggered than is the case with valine. (Methionine is more likely to be correctly identified if the amino acid composition of the peptide is determined prior to sequencing and shown to contain methionine.)

(vi) *Alanine* gives a characteristic sharp red spot right along the top edge of the diethylamine marker and overlapping slightly to the right. There is generally a second red spot (which may even be more intense than the first), slightly above and to the right of the major spot and usually a third, blue, spot in the bottom left-hand corner which has moved almost as far as the diethylamine in the first solvent, but has moved only a few millimetres in the second solvent. This blue spot may be accompanied occasionally by a faint pink spot just above and to its right.

(vii) *Glycine* is difficult to confuse. The primary spot is usually strong and tight up against the right-hand edge of the diethylamine marker. Like alanine it is often accompanied by a further pink spot slightly above and to the right. Also like alanine there may be an additional blue and perhaps pink spot in the bottom left-hand corner, but this time the blue spot barely moves off the origin in the second solvent. After 10 or more cycles, this blue derivative may be the most intense glycine derivative.

(viii) *Serine* even under apparently ideal sequencing conditions, always yields four spots S, S_2, S_3 and S_4 as shown in *Figure 3*. S, the primary spot together with S_4, may be confused with the primary and secondary aspartic acid derivatives (see below), and in practice the best diagnostic spot is S_2 which is almost invariably the strongest under the sequencing conditions described. Serine may be difficult to identify when it occurs after 10 or more cycles, due to build up of background colour on the chromatogram.

(ix) *Glutamic acid* generally gives two secondary spots in addition to the main product, which greatly aid identification. The first of these is red and lies above and to the right of the primary spot. The second is very variable in intensity, blue/grey in colour, and has a mobility similar to the diethylamine marker in the first dimension although it barely moves off the origin in the second dimension.

(x) *Glutamine* usually shows all the glutamic acid spots (presumably due to some deamidation) although the primary glutamic acid spot and its blue secondary spot are often very faint. However, the main glutamine product occurs above and to the right of the glutamic acid red spot. There is often an additional blue spot which has barely moved in the second solvent, but has a mobility mid-way between the glutamic acid and glutamine primary spots in the first dimension.

(xi) *Aspartic acid and asparagine* each produce an equivalent range of spots to glutamic acid and glutamine respectively but with approximately half the mobility in the second solvent.

(xii) *Threonine* yields the three spots shown in *Figure 3*. The primary spot T can be confused with that of glutamic acid/glutamine and T_2 is the best diagnostic feature. It is invariably the most obvious spot, very dense dark blue/grey in colour and with a characteristic tail.

(xiii) *Tyrosine's* red primary spot has a unique position in the bottom left-hand quarter of the chromatogram. It is usually accompanied by a faint red spot immediately above and to its right. Invariably there is also a sharp blue/grey spot immediately to the right of the origin.

(xiv) *Lysine* produces three red spots due to the possession of α- and ϵ-amino groups both of which can be derivatized with DABITC or PITC. Of these, K_3 (equivalent to α-DABTH-ϵ-PTC-lysine) is generally the strongest whilst K and K_2 may together yield a rather indistinct red streak in the second solvent. A very characteristic grey/blue secondary spot almost invariably occurs close to or even overlapping the origin, moving only fractionally in each solvent system. With weak lysine residues, particularly after a large number of degradative cycles when there tends to be a gradual build up of background colour on the chromatogram, this secondary grey/blue spot may be the most useful diagnostic feature indicating lysine.

(xv) *Tryptophan* can sometimes be confused with lysine, or else it is often misassigned since it occurs relatively rarely in proteins and is progressively 'lost' during sequencing. The major red spot is significantly lower than K_3, moving slightly less in the second dimension. There is often a similar secondary spot to that of lysine, close to/just above the origin, but this is generally more purple than grey and is closer to the origin than is the case with lysine. Tryptophan may easily be missed altogether, or misidentified as a weak lysine, unless it occurs within about 10 residues of the N terminus. The presence of tryptophan in the peptide can be confirmed by full amino acid analysis after hydrolysis in barium hydroxide, but may be demonstrated more easily using the Ehrlich reagent (17). Simply apply a sample of the original peptide to filter paper, dry and dip or spray with a freshly prepared 2% solution of 4-dimethylaminobenzaldehyde; 20% HCl in acetone. A violet colour indicates the presence of tryptophan.

(xvi) *Carboxymethylcysteine* (CM-cysteine). The red primary spot for CM-cysteine should not easily be confused with any other except perhaps the secondary tyrosine spot. However, since cysteine occurs relatively rarely, care must be taken to ensure that it is not mistaken for aspartic acid. Such a misidentification may occur particularly when a secondary red spot occurs above and to the left of the primary spot, and there may also be a blue spot directly to the right of the origin in essentially the same position as that of aspartic acid. The overall 'fingerprint' may therefore resemble that of aspartic acid. However, note that the primary red CM-cysteine spot lies to the left of the diethylamine marker whilst the primary red aspartic acid spot lies to the right. One or two additional red spots may occur as a vertical pair close to the origin and are a good aid to correct identification.

(xvii) *Histidine and arginine* can be confused particularly after a large number of degradative cycles have been performed when salt accumulation is likely to have taken place. In general, histidine runs faster and streaks more than arginine in the second solvent. The major spots for both amino acids are accompanied by a very slightly slower-moving blue spot in the first dimension which does not move at all in the second dimension, remaining as a characteristic crescent at the origin (crescent usually more obvious with arginine). Histidine often exhibits a third spot, grey/blue in colour which has a characteristic 'half-moon' appearance. It moves only slightly in the first solvent, when its flat face is the leading edge of the spot, and even less in the second. When it does occur it is absolutely diagnostic of histidine.

4. DANSYL–EDMAN SEQUENCING METHOD

The method described below (18) is very easy to apply and has been successfully used for many years for the elucidation of the sequences of very many peptides and proteins. It has generally been superceded by the DABITC method principally because the former takes somewhat longer, is more expensive with respect to the peptides and cannot easily identify tryptophan or distinguish between the amides and acids. The DNS-amino acids, however, are somewhat easier to identify clearly than is the case for DABTH-amino acids. Most of the strictures mentioned for the DABITC method also apply for the dansyl–Edman procedure.

4.1 **Dansylation** (see also Section 6.3.1 and 6.3.2)

This can be conveniently performed in acid-washed tubes (6 × 50 mm Pyrex; obtained from Corning).

(i) Add 10 μl of 0.2 M $NaHCO_3$ to the dried peptide (1–5 nmol).
(ii) Add 10 μl of DNS–Cl reagent (2.5 mg/ml in acetone).
(iii) Seal with parafilm and incubate for 1 h at 37°C.
(iv) Dry down under vacuum, add 50 μl of 6 M HCl, purge with N_2 and seal the tubes using a hot (e.g. O_2) flame.
(v) Hydrolyse for up to 18 h at 105°C. (Proline- and hydroxyl-containing amino acids are progressively destroyed and if present at the N terminus the hydrolysis times should be reduced to between 6 and 8 h.)
(vi) Open the tubes, dry down *thoroughly* under vacuum, dissolve in 5 μl of ethanol or pyridine and subject to TLC as described in Section 4.3.

4.2 **Edman degradation**

A much larger aliquot of peptide (~100 nmol) can now be subjected to the Edman degradation procedure to remove the residue identified by the dansylation procedure.

(i) Add 20 μl of 50% pyridine to the dried peptide (reaction tubes as in Section 3.1.2).
(ii) Add 100 μl of 5% PITC in pyridine, flush with N_2 and seal the tubes.

(iii) Mix thoroughly and incubate for 1 h at 45°C.
(iv) Dry thoroughly under vacuum/P_2O_5.
(v) Add 200 μl of anhydrous TFA; flush with N_2, seal and incubate for 30 min at 45°C.
(vi) Dry under vacuum.
(vii) Add 150 μl of water and extract three times with 1.5 ml *n*-butyl acetate, mixing (vortexing) very thoroughly each time and separating the layers by centrifugation (1–2 min on a bench centrifuge). Remove the upper layer (butyl acetate) with a fine Pasteur pipette and discard.
(viii) The sample should now be dried under vacuum, re-dissolved in 50% pyridine and an appropriate aliquot taken for the dansylation procedure. The remainder can then be subjected to step (ii) above and the cycle repeated.

4.3 **Chromatography**

Chromatography is carried out in sealed tanks, on polyamide thin-layer sheets, pre-coated on both sides (BDH). It is possible to use 15 × 15 cm sheets, or 7.5 × 7.5 cm or 5 × 5 cm sheets cut from these. The conditions described below refer to 5 × 5 cm sheets. Racks may be simply constructed to allow several sheets to be run together in one chromatography tank.

The solvent systems (prepared freshly) used in this method are:

Solvent 1, 1.5% (v/v) formic acid;
Solvent 2, acetic acid: toluene, 1:9 (v/v);
Solvent 3, methanol:butyl acetate:acetic acid, 40:60:2 (by vol.);
Solvent 4, 0.05 M trisodium phosphate:ethanol 3:1 (v/v).

The sheets may be re-used after washing by immersion in acetone:water:ammonia, 50:40:4 (by vol.) for 1–2 h. The sheets should be thoroughly dried before re-use, and inspected for residual fluorescence.

A standard mixture of DNS-amino acids is also required. This should contain 100 μg/ml each of DNS-proline, leucine (*or* isoleucine), phenylalanine, threonine, glutamic acid, arginine and glycine in acetone or pyridine (tryptophan is an optional extra).

The procedure used is as follows.

(i) Number the sheet in pencil in the top right-hand corner, and mark the origin in the bottom left-hand corner, about 1 cm from either edge. Mark an origin also on the reverse side of the sheet such that the two origins are exactly superimposable.
(ii) Dissolve the sample in 5 μl of pyridine or ethanol and apply to the origins on the thin-layer plate such that 75% is on one side and 25% on the reverse side. Keep the diameter of the spot to less than 2 mm. Use a warm air stream to dry the sample in between applications.
(iii) Apply a 0.5–1 μl aliquot of the DNS-standard mixture to the origin which contains 25% of the sample.
(iv) Place the sheet in the first solvent such that the origin is in the bottom *left-hand* corner. Ensure that the solvent meniscus is well below the origin. Develop until the solvent is about 1 cm from the top of the plate (5–10 min). Remove and

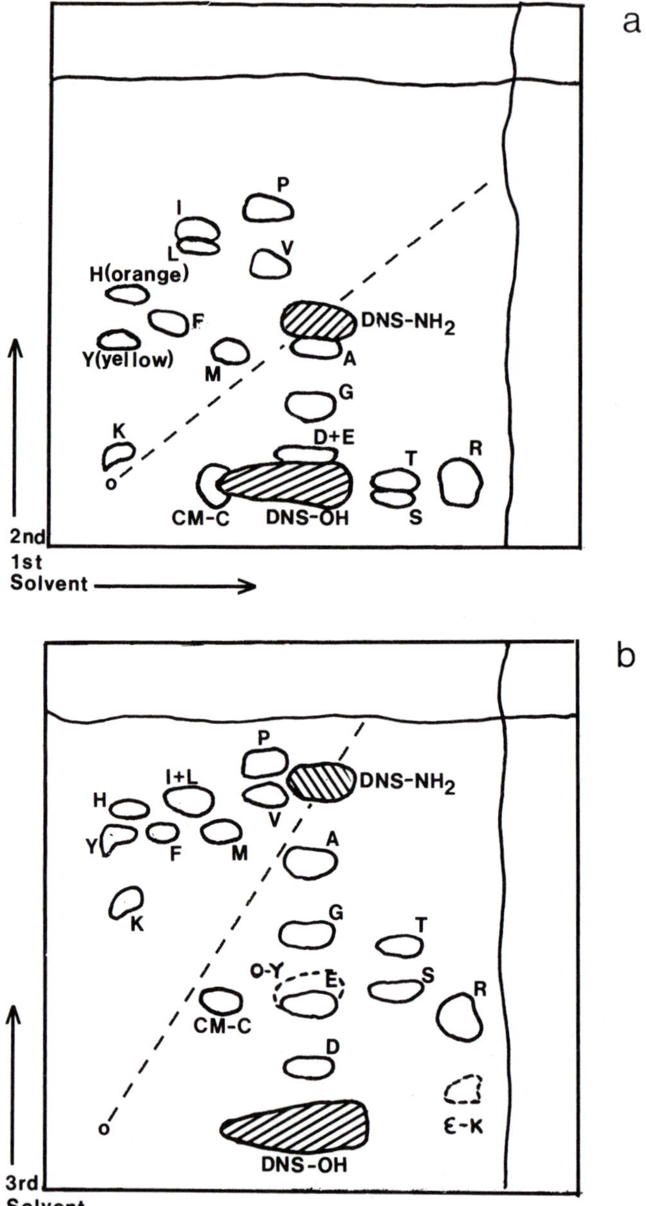

Figure 5. Separation of DNS-amino acids by polyamide sheet chromatography as described in the text. (**a**) After solvents 1 and 2; read spots above the dashed line. (**b**) After solvents 1, 2 and 3; read spots below the dashed line. Amino acids are identified by the single letter code; CM-C is carboxymethylcysteine.

dry in a current of warm air for about 10 min (i.e. until no smell of formic acid remains). View the sheet under UV light.

(v) Carry out chromatography in the second solvent with the sheet at right angles to the direction of movement in the first solvent. Development takes about 10–15 min. Thoroughly dry the sheet and view it under UV light. The separation of DNS-amino acids achieved after development in solvents 1 and 2 is shown in *Figure 5a*. Some DNS-amino acids cannot be unambiguously identified at this stage and further chromatography is required.

(vi) Develop the sheets in the third solvent in the same direction as for solvent 2. Development this time takes 8–10 min. Dry the sheets thoroughly and view them under UV. The separation of DNS-amino acids after solvent 3 is shown in *Figure 5b*. Solvent 4 (again in the second dimension) will resolve DNS-Arg from DNS-ε-Lys.

4.4 **Identification of DNS-amino acids**

(i) The DNS-amino acids are identified by reference to *Figure 5a* and *b* and may be scored on a scale of 1–5 according to apparent relative abundance on the sheet. As a general rule, spots above the diagonal dotted line are read after the second solvent (*Figure 5a*) and only the spots below the diagonal require the third solvent for resolution (*Figure 5b*).

(ii) Identification is aided by reference to the mobility of unknowns plus standards on the reverse side of the sheet ('unknowns' will be less intensely fluorescent than the standards in general; where an 'unknown' is also represented in the standard mix the resultant spot will be more intensely fluorescent than the other standards). The standards are carefully chosen to aid identification of closely running derivatives, for example I/L, S/T, D/E.

(iii) DNS-NH_2 (blue–green) and DNS-OH (blue) are by-products of dansylation, are always present in experimental samples and provide additional marker spots.

(iv) Although most dansyl spots fluoresce green, DNS-His is orange and DNS-Bis-Tyr and DNS-*O*-Tyr are yellow. All are very obvious.

(v) Asparagine and glutamine are deamidated during the hydrolysis stage to give DNS-aspartic acid and DNS-glutamic acid. These should, therefore, be recorded as Asx and Glx respectively, since they may originate from either the acid or the amide. Tryptophan is destroyed.

(vi) Additional spots may occasionally appear due to the occurrence of acid-resistant hydrophobic dipeptides which are subsequently dansylated (e.g. DNS-Ile-Ile) but with experience these rarely cause problems (see ref. 19 for mobilities of dansyl dipeptides).

(vii) The relative mobilities of DNS-Tyr-, -His, -Lys indicated in *Figure 5* are those of the bis-derivatives found when free amino acids are dansylated. If an N-terminal tyrosine, histidine or lysine residue in a peptide or protein is dansylated and subsequently released by acid hydrolysis, the same derivatives will be found, but 'internal' residues will also be dansylated on their side-chains and additional spots (e.g. DNS-*O*-Tyr and DNS-ε-Lys) will then occur on the chromatograms.

5. C-TERMINAL SEQUENCING METHODS

A number of chemical methods for identifying the C-terminal amino acid of proteins and peptides have been described. Of these, hydrazinolysis (20) and tritiation (21,22) have been used with some success. However, no useful chemical method has yet been devised that will enable efficient and convenient sequence determination by *successive* removal of C-terminal residues in a manner analogous to the Edman degradation (however, see ref. 23). Fortunately, sequential enzymic cleavage of C-terminal residues with carboxypeptidases will, in many cases, yield limited sequence data. Essentially, the methods used simply follow the kinetic progress of amino acid release, but care must be exercised for the results can often be complex and subject to misinterpretation because the rates at which different amino acids are released vary considerably. Although the carboxypeptidases all work better on peptides than on proteins, C-terminal sequencing is generally only performed on the intact protein or on the putative C-terminal peptide.

Four carboxypeptidases with different specificities have been used in sequencing studies. The first, carboxypeptidase A (24), has broad specificity but cleaves best at non-polar residues. Aspartate, glutamate, histidine and glycine are cleaved only very slowly, lysine barely, and arginine and proline not at all. The second, carboxypeptidase B (24), cleaves readily at arginine and lysine and not at all at proline. Most other residues are cleaved very slowly, if at all. Since both carboxypeptidase A and B have similar pH optima and, to some extent, complementary specificities, they can profitably be used in combination. A suitable method is as given below.

(i) Dissolve 100 nmol of peptide or denatured protein (reduced and carboxymethylated, citraconylated or performic acid oxidized) in 100 μl of 0.2 M NEM buffer pH 8.5 containing 1 μmol of norleucine (or other internal standard) and add 1.0 nmol of enzyme dissolved in buffer.

(ii) Incubate at 25°C. Remove a 10 μl aliquot at zero time, 5 min and thereafter at 10 min intervals. Shorter incubation periods will probably be required when analysing peptides—try sampling twice as often.

(iii) Add aliquots to an equal volume of 1 M acetic or formic acid to stop the digestion.

(iv) Centrifuge to remove precipitated protein and lyophilize the supernatant.

(v) The released amino acids are best determined by means of an amino acid analyser. Alternatively, they can be identified semi-quantitatively after dansylation (see Section 6.3).

If several free amino acids are already present after 5–10 min, the digest should be repeated using a combination of shorter incubation periods and lower enzyme concentration. In some cases, extended incubations (up to 24 h) and higher enzyme to substrate ratios may be required.

The other two enzymes, carboxypeptidases C (25) and Y (26), are similar in having broad specificity for all residues, including proline, but release aromatic and hydrophobic residues most readily and glycine slowly. They are therefore more likely to give clearer results than the use of carboxypeptidase A and/or B. Use 0.1 M pyridine–acetate buffer pH 5.3–5.5 (enzymes may be irreversibly inactivated above pH 6 and in the incubation and assay conditions described above). If necessary, to aid solubility of proteins, carboxypeptidase A and Y may be used in the presence of up to 6 M urea or 1% SDS with incubations of up to 1 h.

Note: Some commercial preparations of carboxypeptidases may contain significant amounts of free amino acids which should be removed before the enzyme is used. Ammonium sulphate suspensions can be centrifuged and washed with saturated ammonium sulphate before dissolving in buffer. Carboxypeptidase A is insoluble in water.

6. OTHER METHODS

6.1 Reduction and sulphydryl modification

The following procedure converts cysteine to carboxymethylcysteine. The blocking of the free sulphydryl (-SH) group prevents cross-linking of cysteine-containing peptides. The procedure has two other advantages, firstly it denatures the protein making it more susceptible to proteolytic digestion and sometimes more soluble. Secondly, it prevents intrachain cleavages during the Edman degradation reported to occur at unblocked -SH groups.

(i) Dissolve the protein (up to 30 mg) in 3.0 ml of 6 M guanidine−HCl (or 8 M de-ionized urea), 0.6 M Tris−HCl, pH 8.6.
(ii) Add 30 μl of 2-mercaptoethanol (or 5 μl of 1 M dithiothreitol) and incubate under N_2 for 3 h at room temperature. Some proteins may require the presence of 1% SDS and incubation at 30−40°C for 18 h (reduction).
(iii) Add 0.3 ml of *colourless* iodoacetate (268 mg/ml in 0.1 M NaOH) and incubate in the dark for a further 15 min (-SH modification-carboxymethylation).
(iv) Transfer to dialysis tubing and dialyse against 2 × 5 litres of 5 mM NH_4HCO_3 in the dark for 24 h. Alternatively, de-salt on a 20 × 1 cm column of Sephadex G-10 or Biogel P2, in 5 mM NH_4HCO_3.
(v) Lyophilise.

The use of iodoacetate will add new negative charges to the protein. If this is undesirable, 30 mg of iodoacetamide or *N*-iodoethyl-trifluoroacetamide (50-fold molar excess over total thiol) can be added to the reaction mixture in place of the iodoacetate (27). In these cases, no change in charge or the addition of basic groups (-NH_3^+) respectively can be achieved. The two products are *S*-carboxyamidomethyl- and *S*-aminoethylcysteine respectively, the first of which is converted to *S*-carboxymethyl cysteine on acid hydrolysis. Other useful reagents for modifying cysteine residues include methyl iodide (28; methionine is also modified) and 4-vinyl pyridine (29) (see also Chapter 1, Table 10, and Chapter 3, Section 3.4). Note that ethyleneimine is very toxic and should no longer be used; *N*-iodoethyl-trifluoroacetamide is an equivalent substitute (available from Pierce).

6.2 Citraconylation, succinylation and maleylation

Modification of lysine residues by reaction with dicarboxylic anhydrides prevents subsequent cleavage of lysyl peptide bonds with trypsin. This procedure has the added advantage that the modified proteins are generally better substrates for protease digestion, being both denatured and usually more soluble in the buffers used. Citraconic (30), succinic (31) and maleic anhydrides (32) have most commonly been used. Essentially, the native or carboxymethylated protein is reacted with an approximately 50-fold molar excess of the anhydride in the presence of 8 M urea or 6 M guanidine−HCl at pH 8−9. The N-terminal amino group of the protein is also modified.

Singhal and Atassi (33) established that citraconylation is likely to give the most satisfactory results. The following method is based on that of Richardson *et al.* (34).

(i) Dissolve up to 0.5 μmol of protein or peptide in 3 ml of 0.2 M sodium borate buffer, pH 8.5 containing 8 M de-ionized urea or 6 M guanidine−HCl.

(ii) Add citraconic anhydride to give a 40–50 molar excess over amino groups. (Add the anhydride in 10 μl lots at intervals of 15 min.)
(iii) Monitor the pH carefully and maintain at pH 8.3–8.6 by addition of 1 M NaOH.
(iv) Continue the reaction at room temperature for 2 h after the final addition of citraconic anhydride.
(v) Dialyse against 2 × 5 litres of 50 mM NH_4HCO_3, pH 8.1 for 24 h at 2°C or pass through a short de-salting column equilibrated with this buffer.
(vi) To unblock the lysine residues for sequencing, incubate the protein/peptides in 10 mM HCl (pH 2) for 2 h, or 5% formic acid for 8 h at room temperature. All residual acid must then be removed under vacuum before commencing sequencing. The citraconyl group is most easily removed.

6.3 **Semi-quantitative determination of the amino acid composition of peptides**

As stated earlier, it is very useful to have an amino acid composition for each of the peptides being sequenced. This may be valuable in distinguishing between leucine and isoleucine in the sequence and may aid in the identification of other residues. It is most useful in confirming that a peptide has indeed been completely sequenced or that it terminated prematurely due to 'washing-out'. In the absence of an amino acid analyser, a simple, manual, and semi-quantitative estimation of the amino acid composition of peptides may be achieved by total acid hydrolysis, followed by dansylation of the amino acids and identification after separation by TLC. This method is particularly suitable for small peptides where individual amino acids occur only once or twice, but with practise it is possible to 'score' spots of different intensity.

6.3.1 *Hydrolysis*

(i) Transfer the peptide (at least 1–5 nmol) to a 6 × 50 mm glass test-tube. This is best done when HPLC fractions are dissolved in 50% pyridine for sequencing. 10–20 samples can be handled at a time; use numbered hydrolysis tubes etched with a diamond pen.
(ii) Dry the sample *in vacuo* and add 50 μl of 6 M HCl; flush gently with N_2.
(iii) Seal the tube in an oxygen flame by drawing out the neck with forceps and incubate for 20 h at 105°C.
(iv) Open the tube by scratching with a diamond pen, moistening the scratch with water and applying a hot glass rod. Dry the sample *in vacuo* over KOH.
(v) Add 10 μl of 0.2 M $NaHCO_3$ and dry down again (it is important to neutralize residual acid).

6.3.2 *Dansylation*

(i) Mix together 1 vol. of dansyl chloride reagent (5 mg/ml in acetone; store in the dark in a refrigerator) and 1 vol. of water to give the dansyl chloride working solution. This should be clear—cloudiness is due to excess water causing precipitation of dansyl chloride.
(ii) Add 10 μl of dansyl chloride working solution in the re-dried hydrolysate, seal with parafilm and incubate at 45°C for 45 min. The dansylated sample can be analysed directly by TLC as described above (Section 4.3), or it can be stored in the refrigerator until required.

7. ACKNOWLEDGEMENTS

I have drawn freely on the experience of my colleagues Dr M.Richardson and Mr J.Gilroy and on conversation with instructors, participants and friends at the Advanced FEBS Course on Micro-Sequence Analysis of Proteins, Berlin, 1985, in particular Professor J.Salnikow and Dr M.Kimura. However, the interpretations and mistakes are all my own. Finally I should like to thank Ethne Ellis for typing the manuscript.

8. REFERENCES

1. Edman,P. (1950) *Acta Chem. Scand.*, **4**, 283.
2. Edman.,P. (1960) *Ann. N.Y. Acad. Sci.*, **88**, 602.
3. Tarr,G.E. (1977) In *Methods in Enzymology.* Hirs,C.W. and Timasheff,G.N. (eds), Academic Press Inc., New York, Vol. 47, p. 335.
4. Levy,W.P. (1981) In *Methods in Enzymology.* Prestka,S. (ed.), Academic Press Inc., New York, Vol. 79, p. 27.
5. Gray,W.R. and Hartley,B.S. (1963) *Biochem. J.*, **89**, 59.
6. Hartley,B.S. (1970) *Biochem. J.*, **119**, 805.
7. Chang,J.Y., Creaser,E.H. and Bentley,K.W. (1976) *Biochem. J.*, **153**, 607.
8. Chang,J.Y. and Creaser,E.H. (1976) *Biochem. J.*, **157**, 77.
9. Chang,J.Y., Brauer,D. and Wittmann-Liebold,B. (1978) *FEBS Lett.*, **93**, 205.
10. Chang,J.Y. (1979) *Biochim. Biophys. Acta*, **578**, 188.
11. Salnikow,J., Lehmann,A. and Wittmann-Liebold,B. (1981) *Anal. Biochem.*, **117**, 433.
12. Walsh,K.A., Ericsson,L.H., Parmalee,D.C. and Titani,K. (1981) *Annu. Rev. Biochem.*, **50**, 261.
13. Yarwood,A., Richardson,M., Sous-Cavada,B. and Rouge,P. (1985) *FEBS Lett.*, **184**, 104.
14. Lehmann,A. and Wittmann-Liebold,B. (1984) *FEBS Lett.*, **176**, 360.
15. Chang,J. (1983) In *Methods in Enzymology.* Hirs,C.H.W. and Timasheff,S.N. (eds), Academic Press Inc., New York, Vol. 91, p. 455.
16. Yang,C.Y. (1979) *Hoppe-Seyler's Z. Physiol. Chem.*, **360**, 1673.
17. Smith,I. (1953) *Nature*, **171**, 43.
18. Gray,W.R. (1967) In *Methods in Enzymology.* Hirs,C.H.W. (ed.), Academic Press Inc., New York, Vol. 11, p. 469.
19. Sutton,M.R. and Bradshaw,R.A. (1978) *Anal. Biochem.*, **88**, 344.
20. Narita,K., Matsuo,H. and Nakajima,T. (1975) In *Protein Sequence Determination.* Needleman,S.B. (ed.), Springer-Verlag, Berlin, p. 30.
21. Matsuo,H., Fujimoto,Y. and Tatsuno,T. (1966) *Biochem. Biophys. Res. Commun.*, **22**, 69.
22. Ramshaw,J.A.M., Scawen,M.D., Bailey,C.J. and Boulter,D. (1974) *Biochem. J.*, **139**, 583.
23. Parham,M.E. and Londain,G.M. (1978) *Biochem. Biophys. Res. Commun.*, **80**, 1.
24. Ambler,R.P. (1972) In *Methods in Enzymology.* Hirs,C.H.W. and Timasheff,S.N. (eds), Academic Press Inc., New York, Vol. 25, p. 143.
25. Tschesche,H. (1977) In *Methods in Enzymology.* Hirs,C.H.W. and Timasheff,S.N. (eds), Academic Press Inc., New York, Vol. 47, p. 73.
26. Hayashi,R. (1977) In *Methods in Enzymology.* Hirs,C.H.W. and Timasheff,S.N. (eds), Academic Press Inc., New York, Vol. 47, p. 84.
27. Schwartz,W.E., Smith,P.K. and Royer,G.P. (1980) *Anal. Biochem.*, **106**, 43.
28. Rochat,C., Rochat,H. and Edman,P. (1970) *Anal. Biochem.*, **37**, 259.
29. Friedman,M., Krull,L.H. and Cairns,J.F. (1970) *J. Biol. Chem.*, **245**, 3868.
30. Dixon,H.B.F. and Perham,R.N. (1968) *Biochem. J.*, **109**, 312.
31. Klotz,I.M. (1967) In *Methods in Enzymology.* Hirs,C.H.W. (ed.), Academic Press Inc., New York, Vol. 11, p. 576.
32. Butler,P.J.G., Harris,J.I., Hartley,B.S. and Leberman,R. (1969) *Biochem. J.*, **112**, 679.
33. Singhal,R.P. and Atassi,M.Z. (1971) *Biochemistry*, **10**, 1756.
34. Richardson,M., Campos,F.D.A.P., Moreira,R.A., Ainouz,I.L., Begbie,R., Watt,W.B. and Pusztai,A. (1984) *Eur. J. Biochem.*, **144**, 101.

CHAPTER 7

Structure prediction

J.M.THORNTON and W.R.TAYLOR

1. INTRODUCTION

Once a protein has been completely sequenced, it is important to be able to extract the maximum amount of information from it. Experimentally we know that the sequence is sufficient to determine the three-dimensional structure of the protein and how the protein functions, however, we cannot yet predict structure from sequence *ab initio*. In this chapter we aim to provide a simple guide to the most commonly used and powerful methods for analysing protein sequences. These methods are mainly empirically based, and derive from observations and statistical analyses of proteins of known structure. For example, the well-known secondary structure prediction methods of Chou and Fasman (1) and Garnier *et al.* (2) fall into this category.

Currently there are three-dimensional structures for about 300 proteins, including many homologous families. From these data it has become apparent that certain structural features recur in many proteins. These features range from simple structural patterns (e.g. β-hairpin), through functional structures (e.g. the calcium binding α-loop-α structure), to the complex structure of whole proteins. For example, the serine protease structure found in chymotrypsin, trypsin, etc. occurs in many larger proteins and can be identified from the sequence (e.g. tissue plasminogen activator protein). Increasingly, information on such features is being used to improve structure prediction, and since this is the most exciting recent development in the field, we shall describe available patterns and pattern matching techniques in some detail. This will complement the briefer description of the well-used prediction methods reviewed extensively in the past [e.g. Taylor (3)] and that are available in many laboratories.

Figure 1 shows a basic step-wise approach to be followed when analysing a new protein sequence. The first step for any new sequence is to search the sequence data banks to find if the new sequence is homologous to a protein sequence which has already been determined. (Available data banks and search techniques are described in Section 2.) If matches are found for the whole or part of the sequence, the sequences should be carefully aligned and conservation plots drawn to highlight important residues. If the coordinates of a homologous protein are known, this information can be used to build a three-dimensional model structure using computer graphics. This is currently the only method which can predict tertiary structure with any accuracy, and is described in Section 7. If no homologous sequences are found, or if no structures are available, we must resort to *ab initio* prediction methods, starting with secondary structure prediction as described in Section 3. Hydrophobicity profiles are also useful for locating turns, potential antigenic peptides and transmembrane helices (Section 4). The exciting

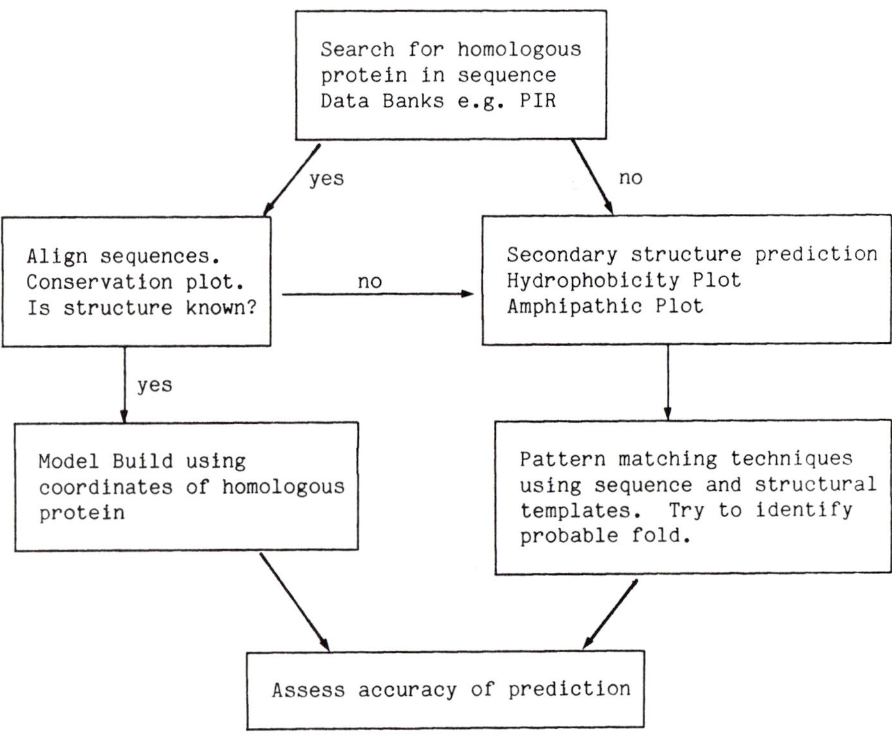

Figure 1. Steps in analysing sequences.

new developments in template recognition and pattern matching, including a reasonably comprehensive list of currently identified sequence templates, are described in Sections 5 and 6. For a single protein sequence these templates can easily be matched by eye and so are very simple to use. At this stage, if there is sufficient evidence in favour of one type of tertiary structure (e.g. β-barrel) it may be possible to model the protein on the basis of a similar structure in the data bank, using the modelling techniques described in Section 7. For example, lipocortin was modelled on the basis of a four helical bundle [Taylor and Geisow (4)]. However, such predictions are still, to some extent, a 'leap into the dark' as it is very difficult to assess their accuracy. As many of the steps towards such a prediction are not automated, they require a good understanding of protein tertiary structure and are not generally recommended. At the end of this chapter we briefly describe possible future developments, including some that attempt to automate these 'intuitive' steps, which we hope will ultimately achieve the prediction of the three-dimensional structure of a protein from its amino acid sequence.

2. SEQUENCE DATA BANKS AND SEARCH TECHNIQUES

There are several large collections of sequence data available. As these have been comprehensively reviewed in a previous volume in this series (5) we will avoid any detailed comparison here. In general, nucleotide sequence libraries are to be avoided as these often present difficulties in choosing the correct reading frame and excising intron sequences. However, they are usually more up-to-date than the protein sequence libraries that are largely translations from the nucleotide sequences. Of the protein sequence libraries, the protein identification resource (PIR) (6) library is recommended especially if installed on a VAX computer as a very useful query program PSQ (protein sequence query) (7) can then be run. This provides facilities to search the library both by name and sequence and includes an increasing amount of information on structural features in the sequence that can be graphically displayed using the VIEW option. PSQ provides two sequence search tools: SCAN searches for exact sequence matches and is very fast (<1 sec) while MATCH allows more flexibility (but no insertions) and takes much longer (many minutes). Although not integral to the PSQ program, other programs are available that directly read the PIR sequences. These include ALIGN which performs a standard Wilbur–Lipman (8) type alignment of two sequences and a similar program SEARCH that does the same over the entire collection of sequences (and takes a long time). For the more ephemeral alignment problems, the program RELATE is useful as this aligns fragments of sequences both within and across specified proteins to aid in the search for partial alignments (e.g. equivalent domains) or internal sequence homologies. [See, for example, its use by White (9).]

The above programs calculate a statistical significance for the alignments they propose based on deviation from the results obtained when shuffled sequences are used. The statistical significance of such measures is notoriously difficult to interpret, especially if the probe sequence has been matched across the entire collection of sequences or the postulated relationship is relatively remote. To a large extent the problem can be avoided by considering patterns of conserved residues, as these should be largely invariant in all members of a family. To identify patterns of conservation requires, of course, that at least two of the sequences are already correctly aligned. If there are more than two known tertiary structures, then even for remotely related sequences, an accurate alignment can generally be determined by structural equivalence that can provide an initial assignment of conserved residues against which other sequences of unknown structure can be aligned [e.g. Taylor (10)]. If there is no structurally determined alignment then the best that can be done is to align the two most closely related sequences and align the others against these. This can be done either by a simple pairwise alignment algorithm [e.g. Taylor (11)] or by one of the many current multiple sequence alignment algorithms. Whatever method is employed the emerging patterns of conservation may, with luck, correspond to some already familiar pattern.

3. SECONDARY STRUCTURE PREDICTIONS

If no homologous sequence is found in the sequence data banks or no structure coordinates are available, then prediction of structure from sequence information alone must be attempted. The first stage is to perform secondary structure prediction, using one of the standard methods to locate α-helices, β-strands, turn and coil regions. Two types of method exist. Most are statistical with parameters derived from observed structures [e.g. Chou and Fasman (1) or the Garnier–Osguthorp–Robson (GOR) method (2)]. In contrast, there are a few methods, notably Lim's (12) that are based on stereochemical principles. These methods are most easily applied using computer

Table 1. Conformational parameters for structure prediction.

	Alpha		*Beta*		*β-Turns*					
Residue	*Pα*		*Pβ*			*P_T*	*f(i)*	*f(i+1)*	*f(i+2)*	*f(i+3)*
ALA	1.42	hα	0.83	iβ	ALA	0.688	0.0494	0.0741	0.0383	0.0494
ARG^+	0.98	iα	0.93	iβ	ARG	0.956	0.0386	0.0734	0.0965	0.0849
ASN	0.67	bα	0.89	iβ	ASN	1.468	0.1478	0.0716	0.1547	0.0762
ASP^-	1.01	Iα	0.54	Bβ	ASP	1.444	0.1338	0.1049	0.1356	0.0687
CYS	0.70	iα	1.19	hβ	CYS	1.013	0.1155	0.0518	0.0478	0.0958
GLN	1.11	hα	1.10	hβ	GLN	0.995	0.0749	0.0629	0.0629	0.1048
GLU^-	1.51	Hα	0.37	Bβ	GLU	0.988	0.0615	0.0945	0.0835	0.0637
GLY	0.57	Bα	0.75	bβ	GLY	1.591	0.0638	0.0745	0.2092	0.1407
HIS^+	1.0	Iα	0.87	iβ	HIS	0.905	0.0939	0.0694	0.0776	0.0367
ILE	1.08	hα	1.60	Hβ	ILE	0.656	0.0640	0.0474	0.0190	0.0711
LEU	1.21	Hα	1.30	hβ	LEU	0.633	0.0555	0.0394	0.0380	0.0613
LYS^+	1.16	hα	0.74	bβ	LYS	0.876	0.0651	0.0961	0.0456	0.0619
MET	1.45	Hα	1.05	hβ	MET	0.788	0.0565	0.0484	0.0403	0.0968
PHE	1.13	hα	1.38	bβ	PHE	0.604	0.0506	0.0281	0.0365	0.0702
PRO	0.57	Bα	0.55	Bβ	PRO	1.281	0.0860	0.1907	0.0628	0.0535
SER	0.77	iα	0.75	bβ	SER	1.357	0.1066	0.1239	0.0942	0.0917
THR	0.83	iα	1.19	hβ	THR	0.979	0.1038	0.0607	0.0511	0.0847
TRP	1.08	hα	1.37	hβ	TRP	0.766	0.0671	0.0268	0.0537	0.0872
TYR	0.69	bα	1.47	Hβ	TYR	0.895	0.0694	0.0578	0.0549	0.0925
VAL	1.06	hα	1.70	Hβ	VAL	0.508	0.0361	0.0476	0.0245	0.0476

Helix and sheet parameters are taken from Chou and Fasman (1). Turn parameters are a revised set derived from 59 high resolution proteins [Wilmot and Thornton (17)]. P_T are the conformational parameters for the the β-turn type; f(i), f(i+1), f(i+2), f(i+3) are bend frequencies in the four positions. The average probability of occurrence for β-turn = 0.3052E-04. A turn is predicted if $<P_T>$ is greater than 0.458E-04.

algorithms which are widely available. [The Biophysics Department at Leeds University has a package which includes all the most popular prediction methods (13).]

3.1 **Methods based on statistics**

3.1.1 *Chou and Fasman method*

The Chou and Fasman method (1) was originally designed to be performed by hand, and since it is quick and simple it is probably the most widely used. Secondary structure propensities for each amino acid type are derived from observed frequencies in proteins of known structure (see *Table 1*). From these propensities the residue types can be classed as strong helix formers (H), weak helix formers (h), Indifferent (I), weak helix breakers (b) and strong helix breakers (B). The residues are similarly classified for β-strand formation. The stages for this method are summarized in *Table 2* and a sample prediction is shown in *Figure 2*.

3.1.2 *Garnier–Osguthorpe–Robson (GOR) method*

In contrast, the GOR method (2) was explicitly designed for computer application and is very elegant to use. For each residue in the sequence, four 'probabilities', P_H, P_E, P_T and P_C ($H = \alpha$, $E = \beta$, T = turn, C = coil) are calculated by summing information from the 17 local residues, $i \pm 8$. This statistical information derived from

Table 2. Stages in secondary structure prediction using Chou and Fasman's method.

1. *Assign* helix and sheet propensity values and symbols.
2. *Search for nucleation sites*
   (a) *Helix*—a 6-residue peptide containing at least four helix formers (H or hα) where Iα counts as ½ hα, and not more than one helix breaker (Bα or bα).
   (b) *Strand*—a 5-residue peptide with at least three Hβ or hβ residues and not more than strand breaker (Bβ or bβ).
3. *Resolve simultaneous helix-sheet assignments.*
   If α and β nucleation sites are predicted for the same residues, calculate the average Pα, Pβ for these residues and compare. The higher probability wins.
4. *Extend helix and strand nucleation sites.*
   The nucleation sites are extended in both directions until the average probability of the end tetra-peptide falls below 1.0. End residues which are breakers should not be included in the secondary structures.
5. *Predict β-turns.*
   For each peptide not assigned to α and β, calculate the turn probability π as the product of four turn propensities P*f* for residues i, $i+1$, $i+2$, $i+3$ (see *Table 1*). NB: amino acids have different P*f* values at different positions within the turn. if $\pi > 4.6 \times 10^{-5}$ then a turn is predicted. This method was originally developed by Lewis *et al.* (14).

Sophistications can be applied at the various levels.

Residue	AA	HELIX PREDICTION Type	Nucl.	Pα	<Pα>		SHEET PREDICTION Type	Nucl.	Pβ	<Pβ>	TURN <Pτ>	TOTAL PREDICTION	X-RAY
1	Lys	h					b						
2	Val	h		1.06			H	β	1.7			β	β
3	Phe	h		1.13		.89	h	β	1.38	1.19		β	
4	Gly	B		.57	.85		b	β	0.75			β	
5	Arg	i	α	.98			i	β	0.93			β	
6	Cys	i	α	.7			h	β	1.19			β	
7	Glu	H	α				B					α	
8	Leu	H	α				h					α	
9	Ala	H	α				i					α	α
10	Ala	H	α				i					α	
11	Ala	H	α				i					α	
12	Met	H	α				h					α	
13	Lys	h	α	1.16			b					α	
14	Arg	i	α	.98			i					α	
15	His	I	α	1.0	.94		i					α	
16	Gly	B	α	.57			b						T
17	Leu	H		1.21			h					T	T
18	Asp	I					B				.83*	T	T
19	Asn	b					i				.45	T	T
20	Tyr	b					H				1.16*	T	T
21	Arg	i					i				.99*	T	T
22	Gly	B					b					T	T
23	Tyr	b					H				.15	T	T
24	Ser	i					b				.21	T	T
25	Leu	H					h				.46	T	
26	Gly	B		.57			b				.67*	T	
27	Asn	b		.67			i	β	.89			β	
28	Trp	h	α	1.08	.87		h	β	1.37			β	
29	Val	h	α	1.06		.99	H	β	1.7	1.2		β	
30	Cys	i	α	0.7			h	β	1.19			β	α
31	Ala	H	α	1.42			i	β	.83			α	
32	Ala	H	α				i					α	
33	Lys	h	α				b					α	

Figure 2. Sample 'Chou–Fasman' prediction for hen egg white lysozyme 1–33. First the α, β types are assigned; then the possible nucleation sites identified. For overlapping α and β sites, the averaging probability is calculated and the largest wins, for example residues 2–5 are predicted β. Then nucleation sites are extended until the average probability <P> is less than 1.0. Turn probabilities are calculated where α,β are not predicted. Turn is predicted if <Pτ> is greater than 0.46 E-04. *Indicates turn predicted. Residue, residue number in sequence; AA, amino acid sequence; type, α, β type assignment; Nucl., nucleation sites for α and β; Pα, Pβ, probability values for α, β for each amino acid taken from *Table 1*; <Pα>, <Pβ>, averaged Pα and Pβ taken over range indicated; <Pτ>, tetrapeptide turn probabilities calculated using f(i), f(i +) . . . parameters in *Table 1*. Values quoted are $\times 10^{-4}$.

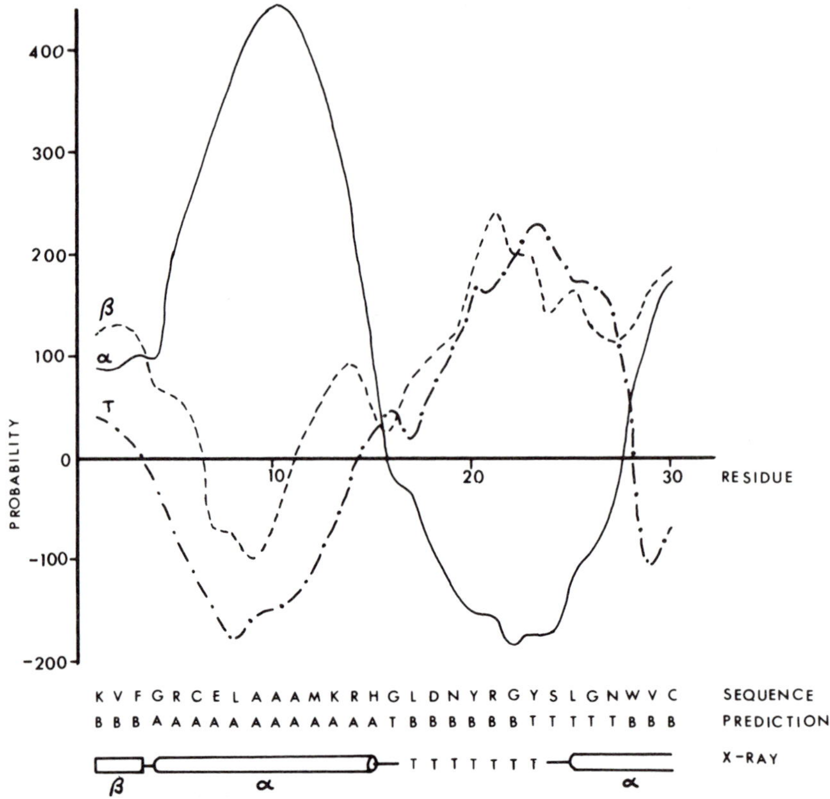

Figure 3. The GOR prediction method (2) applied to the first 30 residues of lysozyme. The graph shows the α, β and turn 'probabilities' calculated by summing over 17 local residues. The predicted state is that with the highest probability. The known X-ray crystallographic secondary structure is also shown.

proteins of known structure is stored in four (17 × 20) matrices, one each for α, β, turn and coil. For any residue the predicted state is the one with the largest probability value. Typical output profiles and the corresponding prediction are shown in *Figure 3*.

3.1.3 *Turn predictors*

The classic β-turn conformations, originally described by Venkatachalan (15), are usually predicted by a four-residue window method using matrices derived from observed amino acid distributions (see *Figure 2* and *Table 2*) (1). From the recent analysis of Richardson (16) five major structural turn types can be recognized: type I, II and their main-chain mirror images, I′ and II′, and type IV. Each type has characteristic ϕ and ψ angles and sequence preferences. However, the standard prediction method groups all these turns together, despite their obvious structural and sequence differences. Using data derived from 59 protein structures (solved to high resolution) Wilmot and Thornton (17) have recently derived separate prediction matrices for the two dominant turn types (I and II). Turn prediction accuracy is significantly increased and, in addition, the turn types can be predicted.

3.2 Methods based on sequence homology

3.2.1 *Homologous peptide methods*

Recent developments in secondary structure prediction have attempted to incorporate longer range interactions and to use the increasing structure data base. For example, Levin *et al.* (18) have used structural information for homologous seven-residue peptide segments to predict secondary structure. For residues 1–7, 2–8, etc. in a test sequence they search for homologous peptides in the structural data base and assign a similarity score using an empirically derived similarity matrix. If the homology is high enough, the structure of each residue in the peptide is assigned to the structure of the homologous peptide, weighted by the similarity score. Once every fragment in the test protein has been compared, the secondary structure attributed to each residue is that which has the highest score. This homologue method had a prediction accuracy of 63% over a limited set of proteins. This is a small but definite improvement over the conventional single residue methods. The approach is similar to that of Kabsch and Sander (unpublished results) and also has affinities with the method of Nishikawa and Ooi (19), both of which attain a similar degree of accuracy.

3.2.2 *Using homologous proteins*

If a family of protein sequences exhibits homology, then these proteins will almost certainly adopt the same basic structure, since structure is conserved much more strongly than amino acid sequence. However, they often give quite different secondary structure predictions. Improvement in accuracy can be obtained by combining the prediction, simply by aligning the sequences and averaging over the predictions [originally recommended by Garnier *et al.* (2)]. For example in the Robson method, the probabilities are simply summed for each aligned residue position and averaged, resulting in a single prediction for the whole family. Improvements in prediction accuracy are again small (~5%) although further improvements can be made by incorporating the known preference for insertions/deletions to occur in coil regions [Zvelebil *et al.* (20)].

3.3 Accuracy

The accuracy of the current prediction algorithms, when applied automatically, is about 55% (21,22). This is not very high and reflects the complex constraints imposed by tertiary structure formation. For example, it has been shown that of pairs of sequence-identical pentapeptides found in proteins of known structure, only about half have the same secondary structure (23). Clearly long range interactions must influence local structure.

This level of accuracy is insufficient to be useful for tertiary structure predictions. For example, using Robson's method on a data set of 53 proteins, the percentage of predicted β-sheet residues which were actually β-sheet in the structure was only 38%. Fortunately, however, there is a correlation between the strength of the prediction (P) and its accuracy. *Figure 4* plots the percent correct prediction against the absolute cut-off value and difference information. Thus where there is a strong helix prediction with high P values, this region is much more likely to be helical than if P were lower. Similarly, if multiple homologous sequences are available, a residue position with a consistent prediction throughout the family is more likely to be correct than one whose predicted

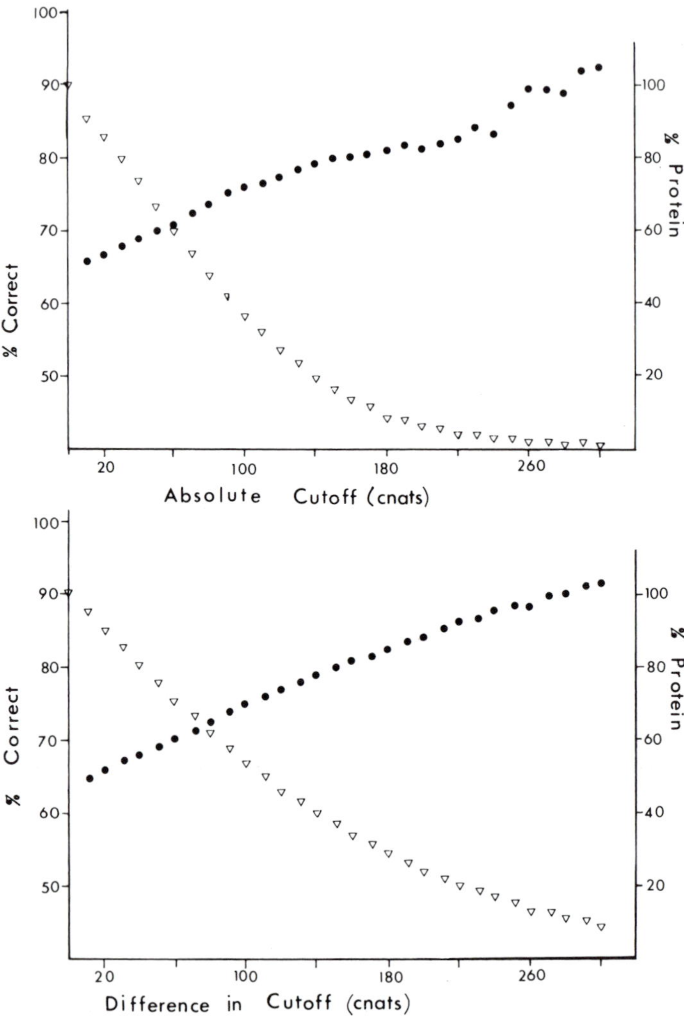

Figure 4. Accuracy of secondary structure prediction. The prediction accuracy (% correct) increases with (**a**) the absolute value of the probability (absolute cutoff) and (**b**) accuracy also increases with the gap between the highest profile and its nearest rival. [The plots derive from the analysis of many proteins (Garratt, Thornton and Taylor, unpublished results).] ● prediction accuracy; ▽ % protein included.

state varies in the different sequences. We are currently using these high accuracy predictions as a core from which the tertiary structure can be modelled.

3.4 Prediction of structural class

We know that tertiary structures can be subdivided into five basic structural classes, all-α, all-β, alternating α/β, $\alpha+\beta$ and 'random' (16,25). Can we use class as the first step towards tertiary structure prediction? From observed compositions for proteins of known structure, the most appropriate partitions are shown in *Figure 5* (26). The predicted structure composition for these proteins varies from reasonably good to

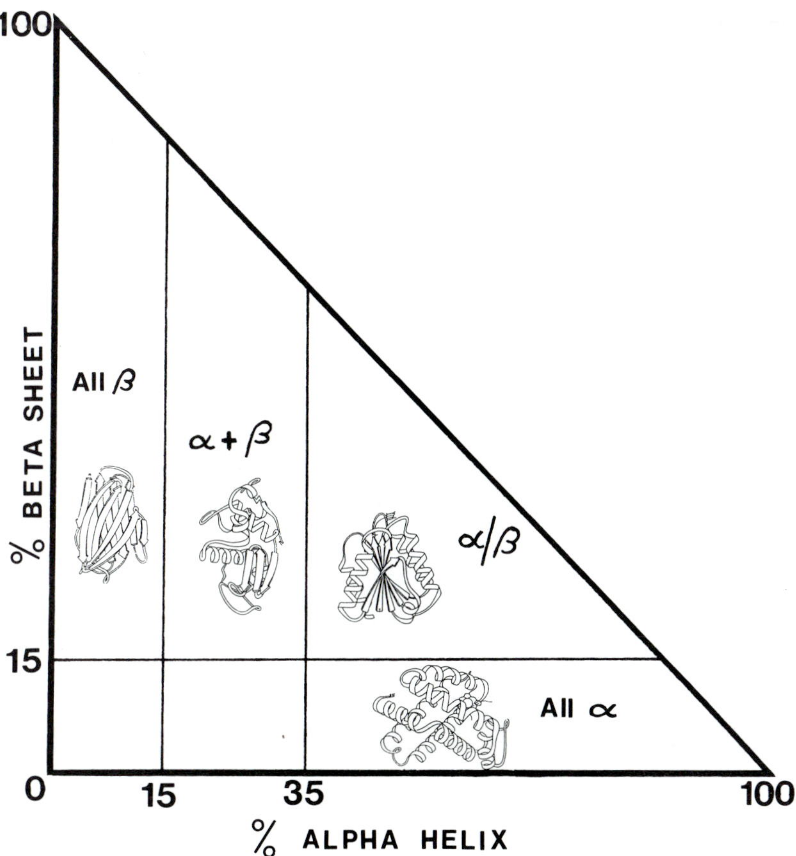

Figure 5. Composition graph of secondary structure content (% α versus % β). Typical members of each protein structural class (in cartoon representation) are plotted. The 15 and 35 percentiles are the best dividers of structural class based on the composition predicted by the GOR method [adapted from (10)].

atrocious, with root-mean-square errors of 16% each. The results of the predicted classes are shown in *Table 3*. Thirty of the 50 proteins were classified correctly with most confusion between the all-β and $\alpha+\beta$ class. Many of the misclassified structures bind relatively large ligands, suggesting that for these structures predictions must be considered with extra caution. Structural class can also provide some functional insight, since intracellular proteins are usually alternating α/β (enzymes) or all-α (non-enzymes). In contrast, extracellular proteins tend to be either all-β (non-enzymes) or $\alpha+\beta$ (enzymes) (27).

4. HYDROPHOBICITY PROFILES

4.1 **Methods**

The 20 different amino acids have been assigned hydrophobicity values, which seek to provide a measure of the partition of that amino acid between a polar (water) and non-polar (protein-interior) environment. There are many different scales of hydro-

Table 3. Comparison of predicted protein class with observed class.

Observed Class	Predicted class: αα	βα	β+α	ββ	*None*
αα	**MBN** **CPV** **C2C** **HMN** **4**	CYT UTG MHB 3			
βα		**SRX** **PGM** **FXN** **PFK** **TIM** **ABP** **LDH** **DFR** **ADK** **PGM** **ADH** **GPD** **B5C** **13**	RHD 1		
β+α	HIP 1	LZM SNS 2	**CAC** **SNI** **CPA** **SSI** **LYZ** **TMV** **ACT** **INS** **SBT** **RNS** **10**	PTI TLN CRN BP2 4	
ββ			CNA PAB CHA PCY 4	**REI** **NXB** **EST** **CRY** **FAB** **APP** **SOD** **7**	
none				FDX RXN 2	

Each protein is represented by its three-letter code as used in the Brookhaven Protein Data Bank. Correct predictions are in **boldface**.
Taken from ref. 26.

phobicity (see ref. 28 for a review) and *Table 4* gives a selection. Although different in absolute values they are broadly comparable in ranking, but individual amino acids may behave differently in the different scales. They are all experimentally derived using different conditions and criteria.

Miller *et al.* (31,34) use the distribution of amino acids between the interior and exterior of protein structures to calculate a scale. This should be the most appropriate scale for the prediction of internal and external residues in proteins. Two consensus scales have been developed [Eisenberg's (33) and Kyte and Doolittle's (32)] which aim to combine the results from the different methods and so 'mitigate the effects of outlying values in any one scale'.

Using the hydrophobicity scales, a profile can be calculated for any sequence by summing and averaging over a window of *n* residues moved along the sequence (e.g. 1–6, 2–7, 3–8 etc.). A window size of six residues is often chosen since this

Table 4. Hydrophobicity scales from the literature.

Amino acid	*Wolfenden et al.*[a]	*Fauchere and Pliska*[b]	*Miller et al.*[c]	*Kyte and Doolittle*[d]	*Eisenberg et al.*[e]
Ile	2.15	1.8	0.74	4.5	0.73
Phe	−0.76	1.79	0.67	2.8	0.61
Val	1.99	1.22	0.61	4.2	0.54
Leu	2.28	1.7	0.65	3.8	0.53
Trp	−5.88	2.25	0.45	−0.9	0.37
Met	−1.48	1.23	0.71	1.9	0.26
Ala	1.94	0.31	0.2	1.8	0.25
Gly	2.39	0	0.06	−0.4	0.16
Cys	−1.24	1.54	0.67	2.5	0.04
Tyr	−6.11	0.96	−0.22	−1.3	0.02
Pro	–	0.72	−0.44	−1.6	−0.07
Thr	−4.88	0.26	−0.26	−0.7	−0.18
Ser	−5.06	−0.04	−0.34	−0.8	−0.26
His	−10.27	0.13	0.04	−3.2	−0.40
Glu	−10.20	−0.64	−1.09	−3.5	−0.62
Asn	−9.68	−0.60	−0.69	−3.5	−0.64
Gln	−9.38	−0.22	−0.74	−3.5	−0.69
Asp	−10.95	−0.77	−0.72	−3.5	−0.72
Lys	−9.52	−0.99	−2.00	−3.9	−1.1
Arg	−19.92	−1.01	−1.34	−4.5	−1.8

[a]Experimental scale derived by measuring ΔG transfer for each amino acid from water to vapour phase. Taken from ref. 29.
[b]Experimental scale derived by measuring the partitioning of the amino acids in octanol/water. Taken from ref. 30.
[c]Expirical scale derived from distribution of amino acids in protein structures. Taken from ref. 31.
[d]Consensus scale derived by considering both experimental partition data and the empirical distribution of amino acids in protein structures. Taken from ref. 32.
[e]Consensus scale as in (d) above. Taken from ref. 33.

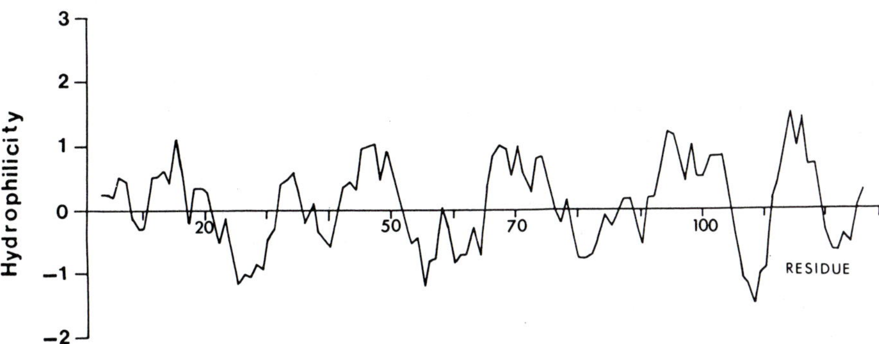

Figure 6. Hydrophobicity plot for hen egg white lysozyme calculated using the Hoop and Wood scale (33).

removes the oscillations associated with the periodic helix or strand and yet retains sufficient local information (see *Figure 6*).

4.2 **Applications**

4.2.1 *Turn predictor*

These hydrophobicity plots can be used to predict exposed and buried regions of a protein, although the correlation between residue accessibility and predicted hydrophobicity is rather poor. For example, the correlation coefficient for myoglobin is only 0.64. Since turns are almost always on the surface of the protein (35) these profiles have been used to locate turns. The method of Rose (36) works by assigning a cut-off value and any region which is more hydrophilic than the cut-off is designated a turn.

4.2.2 *Antigenic peptide predictor*

It is known that peptides, excised from a protein, can be used to raise antibodies which recognize the native structure. Although cross-reactivity is often low and a suitable carrier has to be found, this method can be invaluable in extracting proteins which are found at very low concentrations. Therefore, given the sequence or partial sequence of a protein, it is useful to be able to predict the regions which will be the most effective antigenic peptides. Hopp and Woods (37) used hydrophobic profiles to locate the most hydrophilic segments of the sequence, which were therefore likely to be on the surface of the protein and available for interaction with an antibody (see *Figure 6*). They showed that the peaks in the profile, corresponding to hydrophilic segments were somewhat correlated with antigenic peptides.

Although the results are far from perfect they do provide a starting point for experiment. Karplus and Schulz (38) derived an amino acid mobility scale from crystallographic B-values in proteins of known structure, and a slightly more complex algorithm to predict antigenic peptides. It is based on the observation that mobile segments of proteins are usually exposed and should therefore make good antigenic peptides. Thornton *et al.* (39) have similarly derived a scale of protrusion indices for the amino acids (see *Table 5*) which can be used to derive 'protrusion plots' from the sequence. These linear prediction methods are very simplistic and results do not correlate very well with known structures. If the structure of the protein is known, this is a much

Table 5. Protrusion indices for individual amino acids[a].

Asp	5.77	Ala	4.55
Ser	5.65	Tyr	3.22
Lys	5.57	Val	3.17}
Glu	5.43	His	3.17}
Pro	5.37	Leu	2.93
Gly	5.31	Met	2.89
Gln	4.97	Trp	2.79
Asn	4.82	Cys	2.75
Thr	4.75	Ile	2.71
Arg	4.67	Phe	2.61

[a]Derived for α-carbon ellipsoids calculated for 30 single domain proteins.

more accurate guide to segments which are good candidate peptides, using methods described by Novotny *et al.* (40) or Thornton *et al.* (41).

Two further simple guides may be useful to help in the choice of suitable peptides.

(i) If a family of sequences is available, loop regions are often indicated by greater variation in sequence and insertions/deletions. Such regions will almost certainly lie on the surface of the protein and should make good peptides.

(ii) Sites which are accessible to proteases in the native structure (e.g. trypsin) are usually exposed on the surface in a loop. These sites should also be considered as potential antigenic peptides.

4.2.3 *Transmembrane segments*

In order for a protein to pass through a hydrophobic membrane bilayer, an extended sequence of hydrophobic residues is usually required. Several methods have been described to predict such segments [e.g. (42)], but the major obstacle to improving the accuracy is the limited experimental data defining transmembrane segments.

The minimum length of chain required to span the membrane depends on the conformation of the peptide. Membrane widths are usually quoted as 30−40 Å. Therefore, for a maximally extended chain with α-carbon separations of 3.4 Å, only 9−12 residues are required. However, since all hydrogen bonds should be satisfied within the non-polar bilayer, the α-helical structure is strongly preferred and in this conformation between 20 and 27 hydrophobic residues are needed to span the membrane. Since polar groups are expected at the membrane boundaries, to interact with the polar lipid headgroups, the hydrophobic segment may well be bounded by clusters of charged or polar amino acids.

The simplest method to identify these transmembrane segments is to scan for a peak in a standard hydrophobicity plot (see *Figure 6*). However, since long segments are expected, larger window sizes [e.g. n = 19 by Kyte and Doolittle (32) and n = 21 by Eisenberg (33)] have been used and give clearer plots with less noise. A cut-off is used to identify putative transmembrane segments.

The major problem with this approach is that helices in globular proteins are often incorrectly predicted as membrane-spanning segments, for example, Kyte and Doolittle found their most hydrophobic 19-residue segment in the globular water-soluble lactate dehydrogenase. To avoid this problem Eisenberg (33) suggested that membrane proteins generally include one or two highly hydrophobic segments, termed initiators. Using the consensus scale of hydrophobicities, it is required that these initiators must have an average hydrophobicity over the 21-residue segment of over 0.65 for one or the sum of 1.10 for two. Once an initiator is recognized, then other segments, with average hydrophobicity greater than 0.42 are accepted as membrane-bound segments.

4.3 **Hydrophobic moment**

Eisenberg and co-workers have also developed the concept of the 'hydrophobic moment' (see 33 for review) which can aid the identification of transmembrane and membrane surface-seeking helices. The hydrophobic moment is a measure of the amphipathicity or asymmetry of hydrophobicity of a polypeptide segment. Schiffer and Edmundson (43) first illustrated the importance of amphiphilicity in α-helices by plotting a 'helical

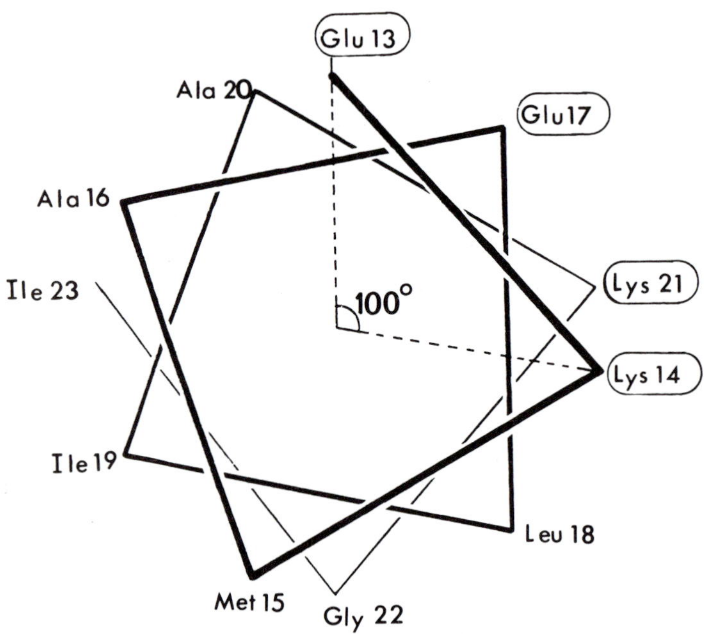

Figure 7. α-Helix viewed along its axis from the amino terminus. The radial distribution of amino acids (helical wheel) shows a distinct (amphipathic) segregation between polar (circled) and non-polar residues. The helix is residues 13–23 of flavodoxin.

wheel'. This is a projection down the helix axis showing the relative locations of side-chains. In globular proteins most helices are distinctly amphipathic (see *Figure 7*) with one side predominantly hydrophobic and the other hydrophilic. In contrast, trans-membrane helices are more uniformly hydrophobic. To quantify this asymmetry, the hydrophobic moment can be calculated for an amino acid sequence and their associated hydrophobicities (see *Table 4*) as the vector sum of the amino acid hydrophobicities. The principle has been generalized to identify any periodicity of hydrophobicity (44).

In a uniformly hydrophobic helix, the hydrophobicities will effectively cancel out, giving a small hydrophobic moment. In contrast, in an amphipathic helix, common in globular proteins, the hydrophobicities will reinforce giving a large value. Thus, from sequence alone, the hydrophobic moment of a putative helix can be calculated and represented as a point on a hydrophobic moment plot. The hydrophobic moment for a peptide segment is plotted against its hydrophobicity. The plot is divided into three regions.

(i) Transmembrane helices, with very high average hydrophobicity and low amphipathicity.
(ii) Helices from globular proteins which are generally less hydrophobic but more amphipathic.
(iii) Helices which seek out the surface of hydrophobicities which are highly amphipathic with high hydrophobic moment.

5. GENERAL SEQUENCE PATTERNS

Sequence patterns associated with specific structural or functional features are being increasingly recognized as more data become available. We have attempted to assemble most of these patterns and, although there is some overlap, we describe the patterns at three different levels.

(i) Structural patterns associated with common secondary, supersecondary and tertiary motifs.
(ii) Patterns associated with specific functional features.
(iii) Patterns associated with homologous protein domains.

These categories will be considered in Sections 5, 6 and 7, respectively.

5.1 **Secondary structure patterns**

The domains of globular proteins have dimensions roughly about 10–40 Å across. To span this distance, a β-strand requires 5–10 residues, while an α-helix needs about 10–25 residues. Thus, for a typical globular protein, sequence patterns associated with elements of secondary structure should be contained within these lengths.

5.1.1 *β-Strand patterns*

If a β-sheet is buried in the hydrophobic core, its strands will be hydrophobic with a preference for branched aliphatic side-chains (Val, Ile, Leu). This very hydrophobic stretch of the strand may be quite short as towards the edge of the sheet more hydrophilic residues will be allowed, especially those with long (hydrophobic) side-chains (particularly Tyr). With these residues the hydrophobic part of the chain can be buried while still allowing the more polar head group to be exposed. If the β-sheet is exposed to solvent on one face then, because alternate residues in the strand contribute to opposite sides of the sheet, a characteristic pattern of alternating hydrophobic and hydrophilic residues is found. β-Strands contain all combinations of the two ideal situations described above as a result of varying degrees of exposure along their length. However, because proteins fall into structural classes there is a greater distinction between the two patterns than might be expected. In proteins with both β and α structure, the sheet is often buried by α-helices on both sides resulting in the characteristic short runs of hydrophobicity while in the all-β proteins two sheets often pack against each other (as in a sandwich) leaving their outer faces exposed and producing an alternating pattern of hydrophobicity.

5.1.2 *α-Helical patterns*

The α-helix is a larger structure than a β-strand and is consequently less likely to be completely buried. Alternating patterns of hydrophobicity are thus generally to be expected in their sequence. These patterns will alternate between hydrophobic and hydrophilic residues with a frequency roughly equal to the periodicity of the helix and frequency analysis methods can be used to identify them in the sequence (see Section 4.3). Alternatively, the sequence can be scanned for basic patterns such as 0--00--0 (where '0' is a hydrophobic residue, and '-' is generally a polar residue). This pattern represents two turns of an α-helix in which the first and the last residues face in the same directon. It is typically contained in the sequences of helices that pack against

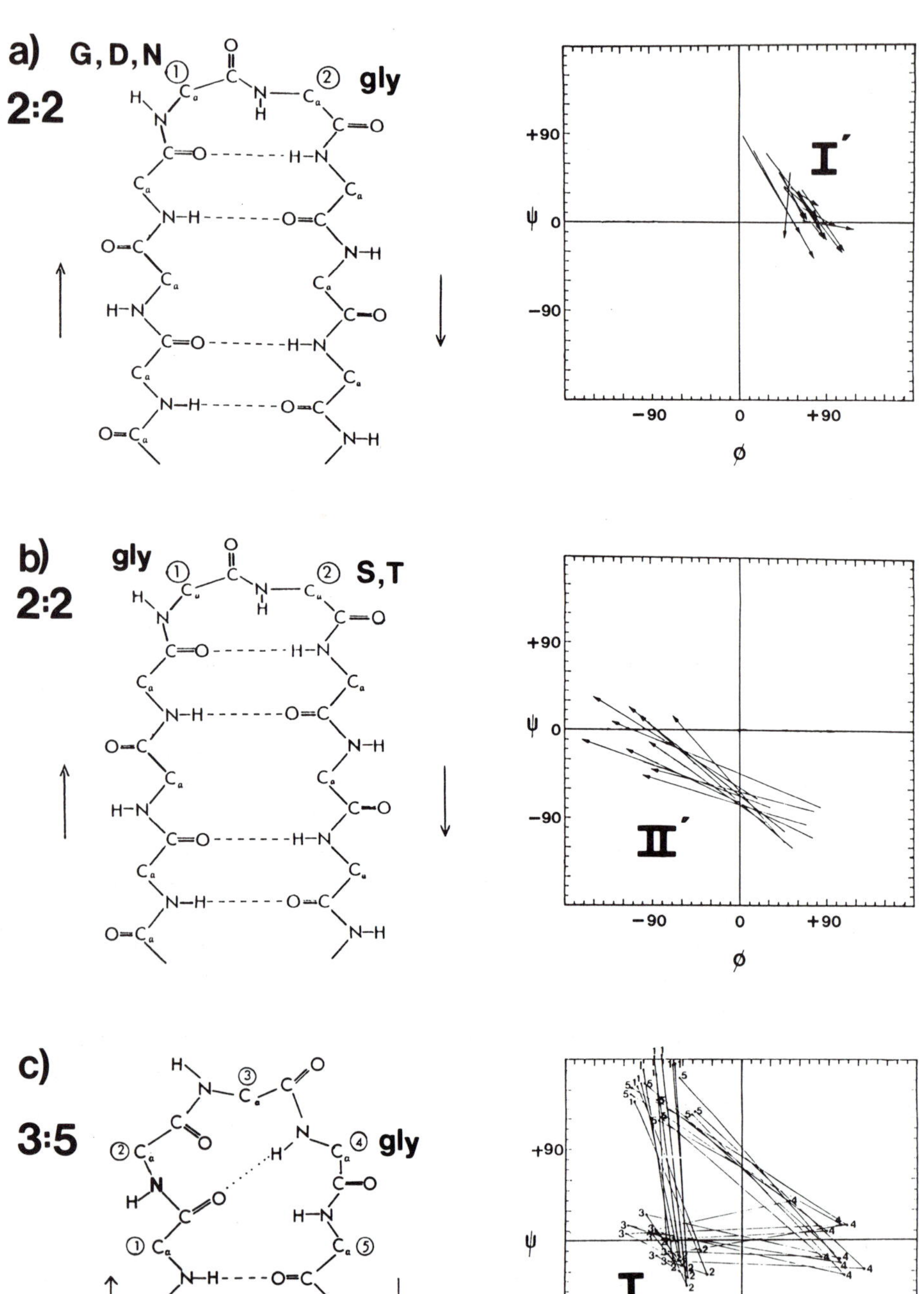
a)
2:2
G, D, N
gly
I´
+90
ψ
0
−90
−90
0
+90
φ
b)
2:2
gly
S, T
II´
c)
3:5
gly
I

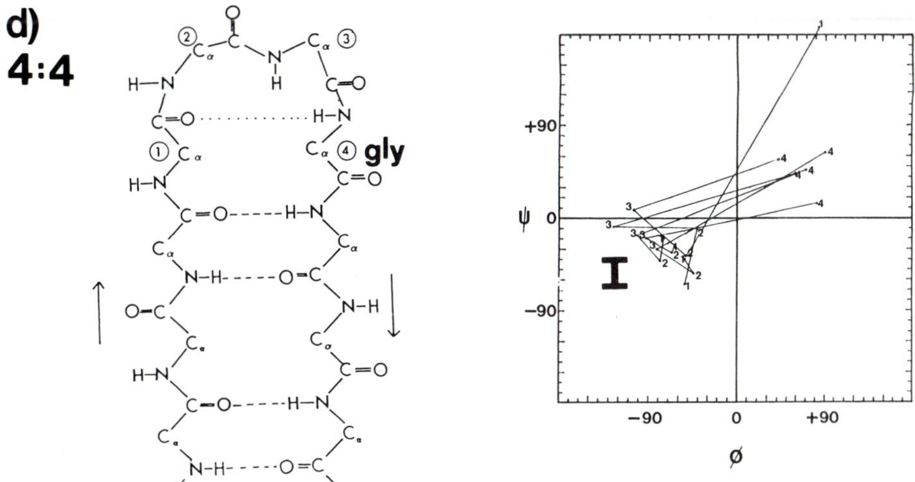

Figure 8. Structural families in β-hairpin loops. For each family a schematic diagram is shown indicating hydrogen bonding, and a dihedral angle plot in which ϕ,ψ values of consecutive residues are joined by a line. (**a**) 2:2 β-hairpin, with a type I′ β-turn; (**b**) 2:2 β-hairpin, with a type II′ β-turn; (**c**) 3:5 β-hairpin with a type I β-turn and a G1 β-bulge; (**d**) 4:4 β-hairpin, with a type I β-turn and a glycine with positive ϕ in the fourth position of the loop. Adapted from ref. 51.

β-sheets in which the positions i and i+7 are well buried and point towards the sheet (45). In this situation it is possible for one of the two middle hydrophobic positions to be more exposed as they tend to splay sidewise and in these positions long chain hydrophilic residues are possible. The basic helical pattern can be extended to longer patterns such as 00--00--00, which describes a twisting hydrophobic face and is characteristic of a helix packing against a β-sheet with a compensating twist (46).

Richmond and Richards (47) identified patterns associated with the packing of α-helices but their analysis was confined to only one protein and is thus of limited generality. However one particular pattern they described involved the close contact of two helices with opposing glycines at the interface with a patch on each helix of the form 00--G--00. This tight interaction has been observed elsewhere and may thus be of greater generality.

5.1.3 *Turn patterns*

As discussed above (Section 4.1) turns are generally hydrophilic. However, more detailed recent analysis is beginning to reveal more specific patterns associated with particular turn types. In the type I turn, the two central residues (i+1, i+2) adopt an α-helical conformation. Proline is particularly common at i+1. The flanking residues, i and i+3, also show some sequence preference, despite variation in their conformational angles (ϕ,ψ). In particular, position i is often Asp or Asn, both of which can form a stabilizing hydrogen bond to $(\text{N-H})_{i+2}$. In contrast, the longer chained Gln and Glu side-chains are infrequent. Similarly, there is a strong preference for Gly at residue i+3, which usually adopts a positive ϕ angle.

Table 6. β-Hairpin loop families.

Type[a]	Conformation[b]	N_{obs}	Loop template
2:2 (I′ β-turn)	$\alpha_L\ \gamma_L$	16	G D G N
2:2 (II′ β-turn)	$\epsilon\ \alpha_R$	10	G S T
3:5 (I β-turn + G1 β-bulge)	$\beta\ \alpha_R\ \gamma_R\ \gamma_L\beta$	10	βTTGβ
4:4	$\alpha_R\ \alpha_R\gamma_R\alpha_L$	5	αTTG

β-Hairpin loop families derived from 107 hairpins in 57 proteins. Forty-one loops fall into one of four structural families (i.e. 38%).
[a]For type see test; [b]conformation is described using nomenclature in ref. 51.
N_{obs} = number observed; in loop template column α, β, T are used to denote residues favouring α-helix, β-strand and type I β-turn structures respectively. Capital letters are one-letter amino acid code.

The dominant sequence characteristics of the type II, I′ and II′ turns are required glycines in conformations sterically forbidden to other residues (although, occasionally an Asp or Asn can be accommodated). In the type II turn, residue $i+2$ adopts an α_L conformation and is almost always glycine. Similarly, in the mirror image type II′ turn, residue $i+1$ is an obligatory glycine. In the type I′ turn, both the $i+1$ and $i+2$ positions are restricted. The $i+1$ residue adopts a left-handed helical conformation and is usually Gly, Asp or Asn; while the $i+2$ residue lies in the $\phi = +90°$, $\psi = 0°$ region and is populated almost exclusively by glycine.

5.2 Supersecondary structure patterns

Certain combinations of secondary structures occur frequently and are referred to as supersecondary structures. Recent analysis of the loop regions in these supersecondary structures has revealed structural and sequence patterns for some of the short loops that connect their component secondary structures. It is likely that specific end-point

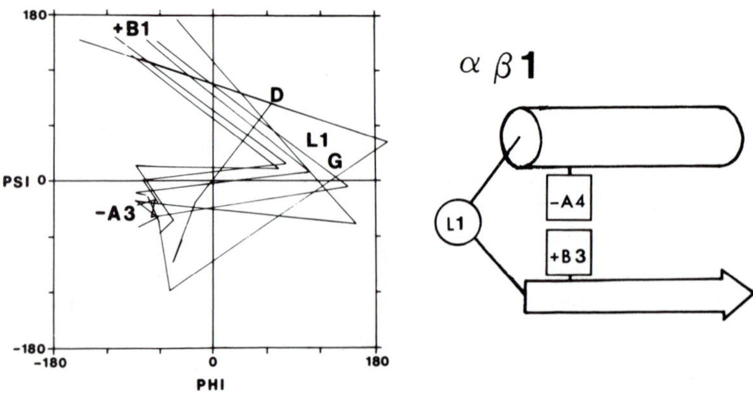

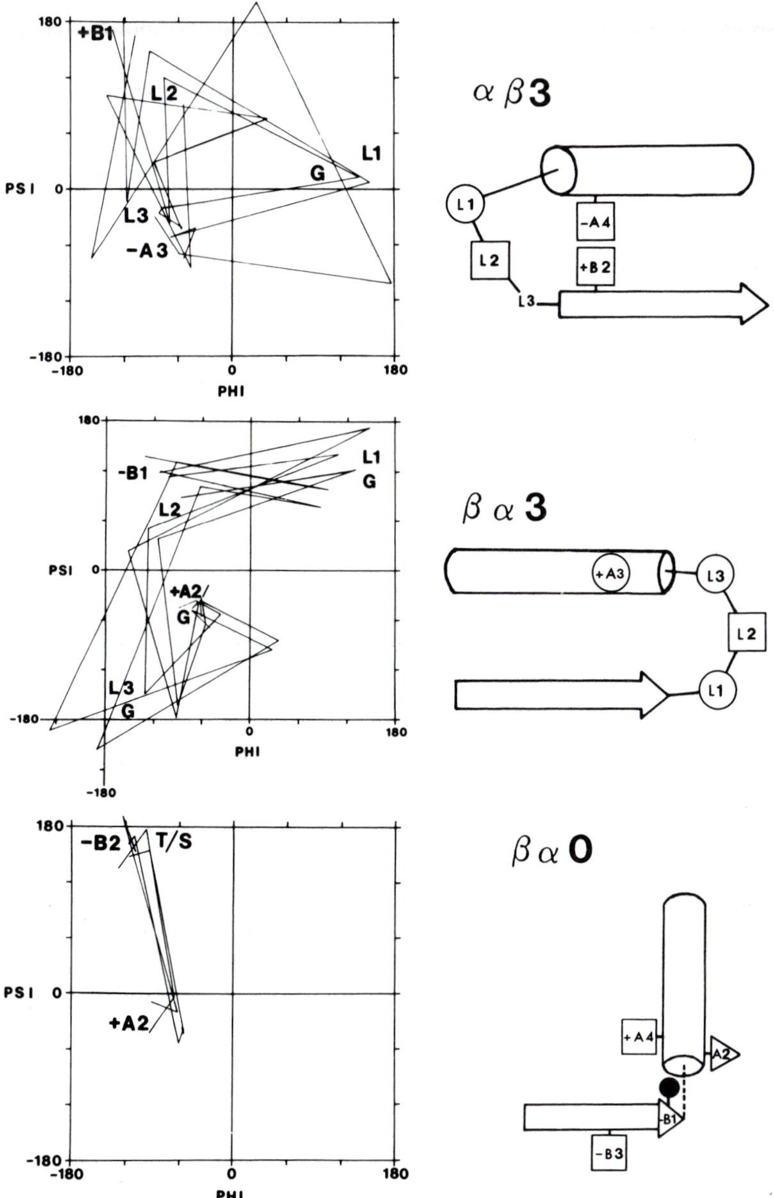

Figure 9. Structural families in $\beta\alpha\beta$ loops. For each family a schematic diagram is drawn, indicating the interacting residues, and the plot of ϕ,ψ angles with successive residues joined. (**a**) $\alpha\beta1$ loop family. ϕ,ψ plot for all members of the family from residues $-$A3 to +B1. The loop residue L1 is a glycine. Hydrophobics at $-$A4 and +B3 pack together. (**b**) $\alpha\beta3$ loop family. ϕ,ψ plot for all members of the family including residues from $-$A3 to +B1. Loop residue L1 is a glycine. Hydrophobics at $-$A4 and +B2 pack together at the helix$-$strand interface. (**c**) $\beta\alpha3$ loop family. ϕ,ψ plot for all members of family from $-$B1 to +A2/A1. Residues L1 and L3 are glycines. The third residue of the helix is glycine or alanine to allow close packing of the helix and strand. Residue L2 is usually hydrophobic and packs into the core. (**d**) $\beta\alpha0$ loop family. ϕ,ψ plot for all members of the family from residue $-$B2 to +A2. The last residue in the β-strand ($-$B1) is serine or threonine, the hydroxyl group of which interacts with the main chain -NH groups of the following three helical residues. Adapted from ref. 49.

distances and geometries lead to the restricted number of loop conformations that are found in unrelated protein structures, and are probably energetically favourable. The structural families found and their sequence templates are summarized below for the three basic supersecondary structures, the $\beta\beta$ hairpin, the $\beta\alpha\beta$ unit and $\alpha\alpha$ loops.

5.2.1 *$\beta\beta$ Hairpin loops*

From an analysis of 107 $\beta\beta$ hairpins in 57 proteins, 41 loops were found in one of four structural families (48). These families are summarized in *Figure 8* and *Table 6*. Two-residue loops are very common in β-hairpins and these are dominated by the rare type I′ and II′ β-turns. Indeed, almost all the I′ β-turns found in the 57 proteins were involved in β-hairpins! These turns both include an obligatory glycine, as do the longer loop structural families. The 3:5 loop includes a type I β-turn and a G1 β-bulge which requires a glycine at position 4 of the loop. The 4:4 loop is similar, with Gly at position 4 but the hydrogen-bonding pattern is distinctive. These four structural families include 38% of all the hairpins in the sample and provide a guide for modelling hairpins.

5.2.2 *$\beta\alpha\beta$ Loops*

The right-handed $\beta\alpha\beta$ unit is ubiquitous in α/β proteins. The strands lie parallel and may be adjacent in the sheet (i.e. hydrogen bonded together) with the helix packing against, and running anti-parallel to, the strands. Despite these structural restrictions, the helix packing is very variable and the loops linking the strands to the helices are longer and more diverse than those found in the $\beta\beta$ hairpins. From the analysis of Edwards *et al.* (49) only 14% of the loops can be categorized (see *Figure 9* and *Table 7*). In the $\alpha\beta 1$ loop a single residue links the helix to the strand. It adopts an α_L conformation and is usually glycine or rarely aspartic acid. The chain reverses direction very sharply leading to tight packing between the helix and strand, with a requirement for hydrophobic residues at $-$A4 and +B3. In $\alpha\beta 3$, the chain reversal is much less abrupt, with three residues in the loop including a glycine at position 1 (with positive ϕ). The second residue is in β-conformation and the third in α_R. Again, there are requirements for hydrophobic residues at $-$A4 and +B2. The $\beta\alpha 3$ loop family includes only nucleotide binding loops with the sequence GxGxxG (see Section 6.1 below). In

Table 7. Loop families in $\beta\alpha\beta$ units.

Type[a]	*Conformation*[b]	N_{obs}	*Loop template*
$\alpha\beta 1$	γ_L	7	ϕ---$\substack{\text{G}\\ \text{D}}$---$\phi$
$\alpha\beta 3$	$\gamma_L\ \beta\alpha_R$	4	ϕ---G$\substack{\text{A}\\ \text{H}}$- -$\phi$
$\beta\alpha 3$	see *Figure 5c*	5	i--ϕ-GϕG -ϕG
$\beta\alpha 0$	–	4	ϕi$\substack{\text{T}\\ \text{S}}$-i-$\phi$

Loop families in $\beta\alpha\beta$ units, derived from 70 $\beta\alpha\beta$ units in 17 α/β proteins. Twenty loops fall in one of four structural families (i.e. 14%).
[a]For type see text; [b]conformation is described using nomenclature shown in ref. 51. N_{obs} = number observed; in loop template column ϕ = hydrophobic residue, i = polar residue, $-$ = any residue, other capital letters are one-letter amino acid code.

the $\beta\alpha0$ group the conformation changes directly from strand to helix without any connecting loop. (The strands in this family lie parallel but are non-adjacent in the sheet.) In the four examples a Ser or Thr is the last residue in the strand and the $-$OH group interacts with the main-chain amide groups of the following three helical residues.

Both $\beta\alpha$ and $\alpha\beta$ turns have been analysed in detail by Efimov (50). The resulting patterns again are not particularly specific as there are no positions that conserve any property other than hydrophobicity. However, the pattern of hydrophobicity and the length of the loops allow the structures to be grouped and patterns identified that may have predictive value.

5.2.3 $\alpha\alpha$ *Loops*

Inspection of the loops connecting two α-helices (which can adopt any relative juxtaposition) by Barlow and Thornton (51) revealed three major structural families (see *Table 8* and *Figure 10*). In the first family the single connecting residue hydrogen bonds to both helices and adopts a β-conformation. The following residue, at the beginning of the second α-helix is often a proline, flanked by two hydrophobic residues. The helical axes subtend an angle of about 60°, making contact through hydrophobic residues at $-$A3 and +A4. The other large $\alpha\alpha$ loop family includes a glycine which adopts an α_L conformation. This conformation is reminiscent of the 3:5 $\beta\beta$ hairpin loop including the $\alpha_R\alpha_L$ β sequence of conformations. This family includes the α-loop-α motif involved in binding DNA (see Section 6.3 below). The remaining family is rather small, but includes a type I β-turn with a proline at position $i+1$. These structural groups include 28% of the 97 loops found in 57 proteins.

The specificity (and hence predictive power) of these sequence templates is very variable, being better for proline-containing templates (since Pro is rare) but low for the glycine patterns. Sequence searches can be made for the above patterns by specifying the patterns as templates including secondary structure probability profiles. For example, a search can be made for two lengths of high β-probability separated by Gly-Gly (for a type I′ β-turn).

6. SPECIFIC FUNCTIONAL PATTERNS

The function of most proteins involves specific interaction with other molecules (including other proteins). Such interactions usually involve only a portion of the protein structure and take the form of a local surface feature that may allow the protein to

Table 8. $\alpha\alpha$ Loops.

Type[a]	*Conformation*[b]	N_{obs}	*Loop template*
0:1	δ/β	13	ϕ-- iPi -ϕ
2:4	γ_R α_L $\beta\beta$	11	ϕ----Gϕ -i--ϕ
3:5	δ α_R γ_R $\beta\beta$	3	ϕ--i- -Pi -i

Structural families in the $\alpha\alpha$ loops, derived from 97 $\alpha\alpha$ loops in 57 proteins. Twenty-seven loops fall into three structural families (i.e. 28%). See legend to *Table 2*.

recognize or be recognized by other proteins, or to bind other molecules ranging from small co-factors and substrates to DNA. Because these binding sites are usually integral to the function of the protein, their structure, and hence amino acid sequence, tends to be conserved through evolution, while other parts of the protein mutate more freely. When comparing relatively remotely related proteins, it can be expected that there will be short regions containing good conservation scattered along the sequence with often little or no conservation between. This pattern of conservation arises as the active or binding sites of proteins are often formed from components contributed from sequentially remote parts of the chain.

In the following sections we will consider the patterns of conserved residues associated with various binding sites and discuss how they can be identified. We will begin with the more general patterns that are common to large families of proteins and progress towards the more specific. It must be remembered, however, that the description of the pattern depends on the degree of homology of the sequences involved. For example, a very specific pattern may be associated with, say, the ATP binding site of the protein kinases which also conforms to a less specific pattern found in all kinases, which may again be associated with a pattern common to all mononucleotide-binding proteins and finally to all nucleotide-binding proteins.

In the following survey we have concentrated on the major families which have known structures, as this allows a better understanding of the mechanics of conservation behind the observed patterns. There are, of course, numerous patterns that are known from sequence data alone and many of these can be associated with specific functions (e.g. enzymic active site residues). There are also many short patterns conserved because they act as a point of recognition and action for other proteins. These are often sites of modification (e.g. glycosylation, cleavage) and generally consist of only a few

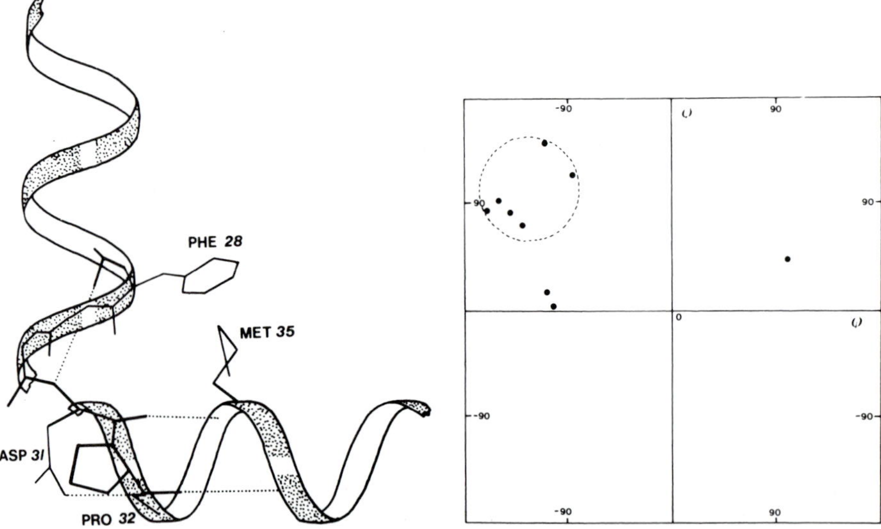

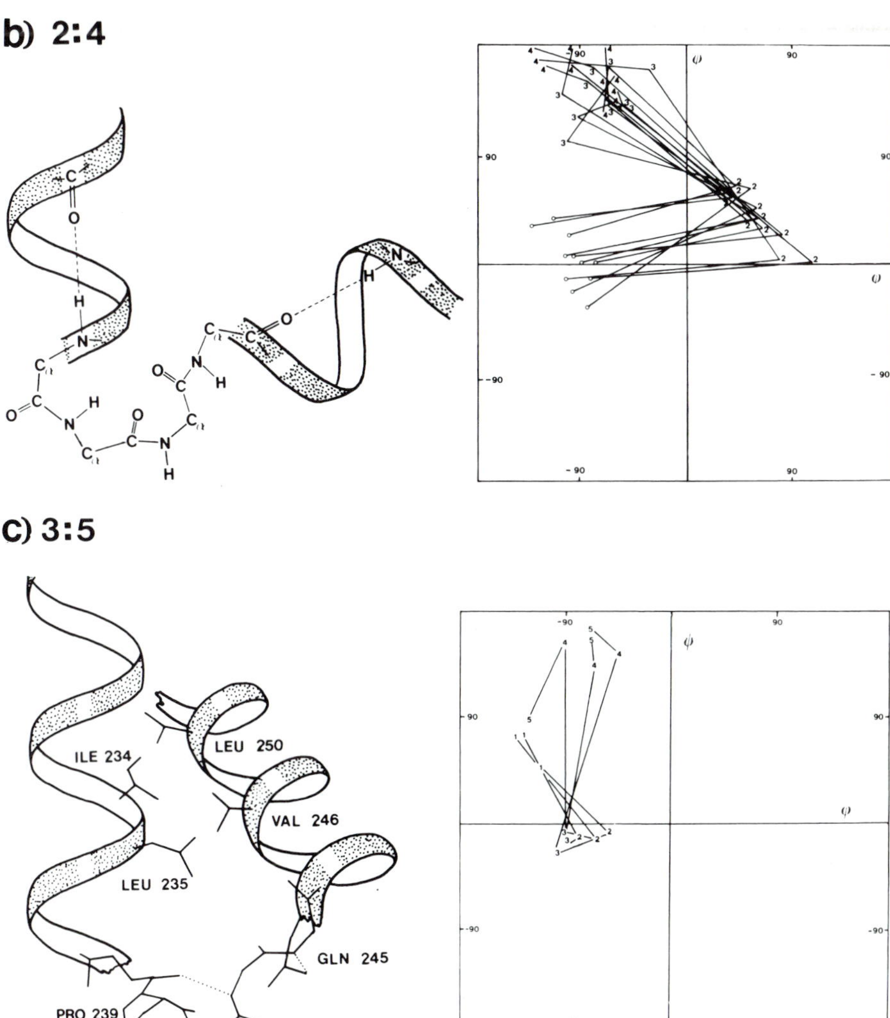

Figure 10. Structural families in $\alpha\alpha$ loops. A schematic diagram and a ϕ,ψ plot for the loop residues in each family are shown. (**a**) 0:1 loop. Only a single residue (L1) has non-alpha ϕ,ψ angles and it contributes hydrogen bonds to both helices. The first residue in the second helix is often a proline. (**b**) 2:4 loop. Residue L2 has positive ϕ and is predominantly glycine. The first residue of the second helix is usually polar and hydrophobics occur at +A4 and −A4. (**c**) 3:5 loop. The proline at residue L2 forms part of a type I β-turn. The helices are closely packed with hydrophobic residues at the interface. The schematic illustrates subtilisin, residues 234−246. Adapted from ref. 51.

residues. They thus have little intrinsic structural interest except as indicators of their accessible surface location. Several examples have been collected in *Table 9*, some of which are reviewed in ref. 52.

6.1 Nucleotide-binding patterns

The large number of proteins that bind nucleotides as a substrate or co-enzyme has attracted much attention to the characterization and identification of nucleotide binding

Table 9. Miscellaneous patterns.

Sequence	*Description/Occurrence*	*Proteins*
GPP	Collagen triple helix (multiple repeats)	collagen, complement Clq.
@..@...	Coiled coil α-helix (multiple repeats)	keratin, desmin tropomyosin, lamins, laminin, fibrinogen, epidermin
Nx−	N-glycosylation site	common in extracellular proteins
KKKRKV	Nuclear protein transit sequence 1	
R##PR	2	
IEGR ·	Factor X protease cleavage site	
GDSGG	Serine protease cleavage site	
FDTGS	Acid protease cleavage site	
#@P@#	Retroviral protease cleavage site	
RPR	Phagocytosis uptake	
RGDS	Fibronectin cell adhesion site	
H<3-4>H	Cu^{2+} binding site	
#ST#K	Penicillin binding	
#S#TK	site	
PxPx.$	Ser or Thr phosphorylation site	

In this and the following tables groups of amino acids are represented in the following convention:
Amino acid types are represented by the standard one-letter code:
A = alanine; B = aspartic or asparagine; C = cysteine; D = aspartic; E = glutamic; F = phenylalanine; G = glycine; H = histidine; I = isoleucine; K = lysine; L = leucine; M = methionine; N = asparagine; P = proline; Q = glutamine; R = arginine; S = serine; T = threonine; V = valine; W = tryptophan; Y = tyrosine; Z = glutamic or glutamine.
Where the position in a pattern represents just one amino acid, the letter is printed bold. If the position conserves predominantly one amino acid, its code letter is printed normally. Other positions use the following groups. Residues that occasionally are included in the group are indicated in parentheses after the main members of the group.

Symbol = Group	Name
§ = A G	
£ = I L	
¥ = Y F	
$ = S T	
R = K R	
+ = K R H (N Q)	positive
− = D E (S T)	negative
± = K R D E (H)	charged
β = L I V (A)	β-branched aliphatic
@ = L I V F M (A P W G)	strongly hydrophobic
# = L I V F M T A P Y W G (K H)	generally hydrophobic
P = S N D E Q R K (H T C Y W)	polar
s = G A S C V (D N T L I)	small
. = no conservation	

If a position in a pattern is confined to the main group then the symbol is printed bold.
Insertions and deletions are indicated by specifying the range of missing residues in angle brackets, for example <2−6>. (A single value simply indicates the length of sequence not represented.)

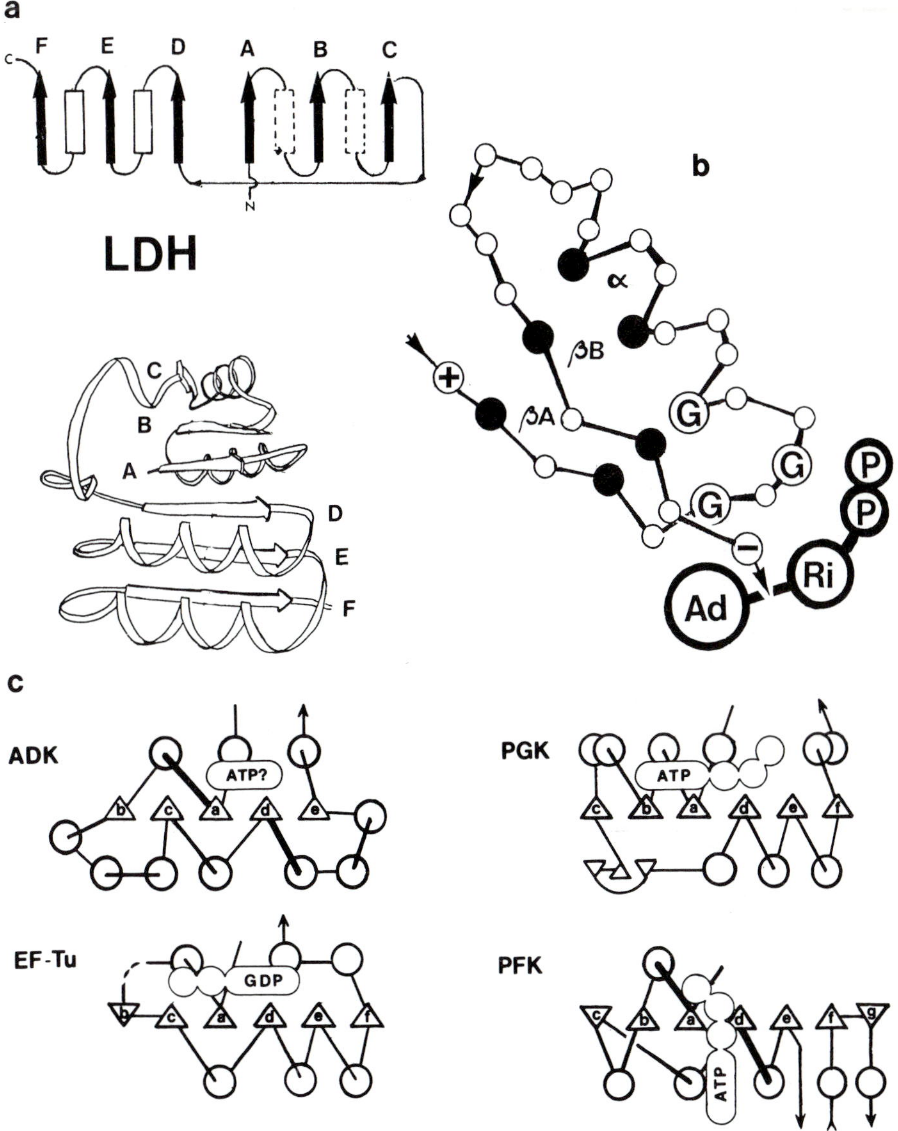

Figure 11. (**a**) Schematic representations of the nucleotide binding domain in lactate dehydrogenase (LDH). β-strands (arrows) are sequentially lettered (A−F). The nucleotide generally binds along the C-terminal ends of the β-strands. (**b**) Simplified representation (connected α-carbons) of a dinucleotide binding βαβ unit (adapted from ref. 59). Conserved hydrophobic residues = @ charged residues = + and − and glycines = G. The bound adenyl part of the nucleotide (F/NAD) is schematically represented as Ad = adenine, Ri = ribose sugar, P = phosphate. (**c**) Symbolic representation of some mononucleotide (ATP/ADP, GTP/GDP) binding domains. The domains are viewed edge on (from the plane of the page in **a**) with β-strands represented as triangles and α-helices as circles. The approximate location of the bound nucleotide is shown. Loops in which conserved residue patterns are found are drawn bold. The proteins are: ADK, adenylate kinase; PGK, phosphoglycerate kinase; PFK, phosphofructokinase and EF-Tu, ribosomal elongation factor Tu.

sites. Because nucleotides are the principal energy currency of the cell the proteins involved are almost all intracellular enzymes, or are on oxidation reduction pathways. Despite the variety of function found among these proteins, most contain a distinct domain responsible for binding the nucleotide. This nucleotide binding domain is almost always of the β/α structural class (see Section 5.2.2). Besides the parallel stranded β-sheet, characteristic of this class, most nucleotide binding domains share a very basic chain fold topology. Considering lactate dehydrogenase [Rossmann *et al.* (53)] as an ideal example of the fold, the chain begins in the middle of the sheet and winds through two $\beta\alpha\beta$ units to the sheet edge. From there it 'jumps' back to a position in the sheet adjacent to, but on the far side of the first strand (referred to below as the D-strand) and then continues to wind towards the opposite side of the sheet. The nucleotide invariably binds along the C-terminal edge of the sheet with the loops following the first (A) and fourth (D) strands central to the site (54) (see *Figure 11*).

Because of the location of these loops in the binding site, their sequence is generally well conserved, giving rise to a pattern characteristic of nucleotide binding sites of two conserved hydrophilic regions separated by roughly 50–100 residues. In the following discussion both these regions will be considered in detail separately and referred to as the N-terminal and C-terminal patterns. In an isolated domain, the N-terminal pattern should be near the N terminus (but not too close, as it must be preceded by a β-strand) while the more C-terminal pattern should occur roughly in the middle of the domain (see *Table 10* for details).

6.1.1 *N-Terminal nucleotide binding patterns*

(i) Dinucleotide binding. Analysis of the sequences of domains that bind dinucleotides (e.g. NAD, FAD), including mainly dehydrogenases, revealed residue patterns associated with the loop following the first β-strand in the domain. A sequence pattern, composed principally of glycines, was recognized by Wierenga and Hol (55) to be associated with the binding of a dinucleotide at this loop. Three glycines lie in the sequence GxGxxG (where x is any residue). Their occurrence in this pattern was rationalized with the first glycine (by its lack of side chain) forming a binding pocket for the adenine group and the second allowing the phosphates of the co-enzyme to interact closely with the dipole of the α-helix (56) while both help to maintain the turn potential of the β-turn-α structure in which they occur. The third glycine is in the α-helix and forms part of the β/α contact allowing a tight turn to be made. Although dominated by glycine, the pattern, or fingerprint, also comprises hydrophobic positions characteristic of a buried β-strand preceding the first glycine and similar positions characteristic of an amphipathic α-helix following the glycine. After a short helix the chain returns to the β-sheet, again with typically hydrophobic residues, and the pattern ends with a conserved aspartic acid which begins a loop lying adjacent to the glycine-rich loop and hydrogen-bonds with the nucleotide ribose hydroxyls (see *Figure 11b*).

The structural implication of this conserved pattern was used by Wierenga and Hol (55) to interpret the sequence change in the GTP binding site of p21 to its oncogene variant in which one of the key glycine positions is a valine and probably prevents nucleotide binding. It was similarly used by Sternberg and Taylor (57) to predict the ATP binding site of the p60 *src* oncogene product.

Table 10. Nucleotide binding patterns.

(a) N-terminal patterns

Occurrence: near N terminus of nucleotide binding domains. βa is never on the edge of the β-sheet.
(i) dinucleotide binding (FAD, NAD)
Proteins: dehydrogenases and reductases but also oxidases and hydroxylases.

```
 βa               α1                        βb
βββββββ  αααααααααααααααα    <1−5>   βββββββ
+ # . # .G.G..G... # .. # ...           . # . # . −
```

Notes: the final position (−) can be positive in proteins where a phosphorylated co-enzyme is bound.
(ii) Mononucleotide binding (ATP/ADP, GTP/GDP)
Proteins: kinases, ATPases, *ras* type oncogene products, G-proteins, ribosomal elongation and initiation factors, tubulins and myosin (light chain).

```
 βα                        α1                βb (usually edge strand)
βββββββ              ααααααααααααα           ββββββ
+ +@@@@GsGssGK$.. # # .@@..  <1−7>
```

Notes: the first few positions generally contain a mixture of positive and hydrophobic residues.

(b) C-terminal patterns

(i) Adenyl binding (ATP/ADP)
Occurrence: middle of nucleotide binding domains (50 − 100 residues after pattern 1a. βd lies adjacent to strand βa (of pattern 1b) in the β-sheet.
Proteins: kinases, ATPases

```
                 βd        ATP-binding
αααααααα    ββββββ     loop
# # . # # +p.G.p # @@@ # DDD
```

Notes: positions 'D' should contain at least one aspartic residue. Glycine is common in the other two positions
(ii) Guanyl binding (GTP/GDP)
Proteins: Elongation factor-Tu, *ras* oncogene products, G-proteins and transducins.
Occurrence: middle towards end of domain. β-strands βc, βd and βe lie adjacent in the sheet with βc next to strand βa (in pattern 1b).

```
βc G-binding            α βd α             βe           G-binding
loop                                    ββββββββ     loop
# @DssG                 < 45−60 >       @@@@ #       NK.D
```

Notes: between βa (GxGxxG nucleotide binding pattern) and βc there is no conservation across the super-family and the alignment around βc is ambiguous.

(c) Cyclic nucleotide binding (cAMP, cGMP)

Proteins: catabolite activator protein (CAP), cAMP-dependent kinase regulatory domains (cAMP-DKRD) and cGMP-dependent kinase (cGMP-DK).
Occurrence: in single or duplicated all-β domain at N terminus (CAP), C terminus (cAMP-DKRD) or middle (cGMP-DK).

```
 β2          β3             β4           β5            β6                     β7        β8
ββββββ  ββββββββββ  βββββββ                                            ββββββββββ
@@.QG-ps-p@¥ # I.pGp # s@@ # p    <13−25>   GELsL@    <2−3>   pPRsA. # .A
```

Notes: the domain in CAP links to the DNA binding domain through a long helix following β8 which also interacts with cAMP.

Subsequent, more extensive, analysis (58,59) confirmed the Wierenga and Hol pattern for the dinucleotide binding proteins of known structure but found that in other closely homologous sequences only the central glycine of the GxGxxG pattern is invariant, the other positions were seen to allow small residues. An exception was also found in the conservation of the C-terminal aspartic which occasionally was glutamic. Interestingly, when the molecule bound was NADP, a positively charged side-chain was found to replace the Asp and interact with the 2′ sugar phosphate.

Wierenga *et al.* (60) noted that only the adenine moiety adopted a constant position with variation in location of the flavin (in FAD) and nicotinimide (in NAD). Consequently they identified their fingerprint as essentially an adenyl binding pattern which explains its similarity to the mononucleotide (e.g. ATP) binding patterns discussed below.

(ii) *Mononucleotide binding.* A sequence pattern similar to that of Wierenga and Hol has been found by Walker *et al.* (61) to be conserved in some kinase and ATPase sequences. This pattern, which in the loop region is characterized by the sequence GxxGxGK, is superficially very like the Wierenga–Hol pattern. However, in the kinase of known structure included in Walker *et al.*'s list [adenylate kinase (62)] the loop has a significantly different structure from that described in the dehydrogenases and reductases by Wierenga *et al.* (60). Although the kinase/ATPase glycine pattern is associated with a loop that is topologically equivalent to the loop containing the dehydrogenase/reductase pattern (between the first β-strand and following α-helix of the nucleotide binding domain) the final xxG of the GxGxxG dehydrogenase pattern is part of the following α-helix, while in adenylate kinase all the glycines are contained in an extended loop. The interaction of this loop with ATP bound by adenylate kinase (63) differs in detail from nucleotides bound by dehydrogenases and reductases. Thus, all that the two glycine patterns have in common is that of a glycine-rich loop interacting principally with the nucleotide phosphates.

When GTP is bound, the patterns of glycines is more similar to that found in the dinucleotide binding domains [enabling Wierenga and Hol to predict the binding site of the GTP-binding protein p21 from the structure of the NADP-binding *p*-hydroxybenzolate hydroxylase (55)]. The partial crystallographic structure of the GTP binding domain of elongation factor EF-Tu (64), which is homologous to the *ras* oncogene products and other G-proteins (including p21) allows this prediction to be confirmed as the structure was determined with GDP bound.

6.1.2 *C-Terminal nucleotide binding patterns*

The patterns associated with the more C-terminal region of conservation (in the middle of the domain, ~100 residues after the above patterns) are more variable between the different nucleotide binding families, with different patterns found in mononucleotide binding and almost none associated with dinucleotide binding proteins.

(i) *Dinucleotide binding.* An invariant glycine is found towards the end of the loop following the fourth (D) strand in the dehydrogenases but there is no apparent structural or functional reason for this isolated conservation.

(ii) *Adenyl mononucleotide binding.* Walker *et al.* (61) noted a region of conservation about 50–100 residues after the N-terminal conserved glycine pattern, which in the two kinases of known structure included in their alignment (adenylate kinase and phosphofructokinase) forms an $\alpha\beta$ structure. The pattern is characteristic of the secondary structure in which it occurs, consisting of OxxR at the C terminus of the α-helix (positive charged residues are favoured at the C termini of helices), through a conserved glycine in the loop, to a very hydrophobic stretch associated with the buried β-strand and finally to a negatively charged region in the following loop that is involved in nucleotide binding (possibly to the Mg^{2+} of an Mg–ATP complex). The pattern has no equivalent in the dehydrogenases beyond a general correspondence of hydrophobic and turn-favouring residues expected from their equivalent secondary structures (see *Table 10b*).

The pattern identified by Walker *et al.* in the mitochondrial F1-ATPase was sufficiently close to adenylate kinase for a model to be proposed for the F1-ATPases based on the known structure of adenylate kinase (65). Not all kinases or ATPases contain the patterns described by Walker *et al.* and in the various cation-transport ATPases associated with the cell membrane, the N-terminal ATP-binding pattern is not found. However, a reasonable match to the more C-terminal pattern allowed Taylor and Green (66) to propose a model, again based on adenylate kinase, for the nucleotide binding domain of those ATPases.

(iii) *Guanyl mononucleotide binding.* An extensive sequence alignment has been determined over several families of guanine nucleotide-binding proteins, including the family of G-proteins, elongation factors and the *ras* oncogene products. Besides the characteristic kinase-like glycine pattern near the N terminus, there are some regions of sequence conserved in almost all members of this superfamily. These consist of a DxxG found some 100–200 residues after the Gly-rich pattern followed 50–60 residues later by a strongly hydrophobic run and the sequence NKxD (67–69) (see *Table 10b*).

Light has recently been shed on the significance of these patterns by the solution of a crystal structure of a fragment of elongation factor Tu (EF-Tu) with Mg^{2+} GDP bound (64). The N-terminal glycine pattern binds the nucleotide phosphates, as expected, in a loop following the first β-strand (A). After a region of great sequence variation between the families, the chain returns again to the nucleotide binding site where the conserved DxxG sequence interacts with the Mg^{2+} at the end of strand C. The nucleotide is bound differently from the ATPase/kinase site and instead of interacting with the loop at the end of the D-strand (the loop normally conserved in ATP binding), interacts with the loop following strand E where the conserved NKxD sequence is located.

6.1.3 *Cyclic nucleotide binding*

The catabolite activator protein (CAP) of *Escherichia coli* binds both DNA (see below) and cAMP. This is the only known structure in which a cyclic nucleotide is seen bound by protein. The structure of the domain to which it binds is completely different from any of those discussed above and is an all-β structure (with the topology of a jelly-roll barrel).

There are some sequences that are homologous to the cAMP binding domain of CAP and although together these do not constitute an extensive family, the recurrence of the domain is of interest because of the central role of cAMP in cell regulation. The alignment proposed by Weber *et al.* (70) relates the CAP domain to the sequences of the regulatory subunits of the cAMP-dependent kinases. These regulatory subunits contain an internal repeat of the domain which is, interestingly, also seen in the cGMP-dependent kinase where two regulatory (cGMP binding) domains precede the catalytic domain.

The cAMP on CAP is held mainly between two β-hairpins (strands 2–3 and 6–7) and the long helix bridging to the DNA binding domain. When aligned with the other sequences two residues (QG) in the turn between strands 2 and 3 are conserved, probably for structural reasons as they do not make contact with the cAMP. The main region of conservation is found from the end of strand 6 into strand 7 with surprisingly little conservation found in the adjacent helix (see *Table 10c*).

6.2 Calcium-binding patterns

Many proteins bind calcium. In some of these the Ca^{2+} ion binds transiently and is involved in conformational changes in the protein structure while in others the Ca^{2+} appears simply to fulfil a more passive structural role. In the latter class the ion is generally held in an extended loop on the protein surface which, for the want of better analysis, can only be described as variable in structure from protein to protein. No characteristic patterns are apparent in the sequence beyond the individual protein families, apart from a bias towards negatively charged residues in the loop. In some proteins glutamic acid residues in the Ca^{2+} binding site are post-translationally modified to γ-carboxy glutamic acid which can bind Ca^{2+} tightly.

There is, however, a large extended family of proteins in which Ca^{2+} plays a central role in modulating structure and behaviour. These proteins are generally message propagators or transducers as they in turn modulate the behaviour of other proteins. This superfamily which includes troponin, parvalbumin and the intestinal Ca^{2+}-binding protein is often typified by calmodulin.

6.2.1 *EF-hand*

The first member of the family to be structurally determined was parvalbumin (71) which was found to be a small protein consisting of six α-helices. In this structure two Ca^{2+} ions are bound in extended loops connecting the two final pairs of helices (C-D and E-F). These helical pairs have almost identical structures and are related by an approximate 2-fold symmetry which is reflected in an internal sequence homology. The parvalbumin structure thus immediately identified a simple helix-loop-helix Ca^{2+} binding motif. The motif takes its name from the final helix pair in parvalbumin and its structural similarity to a particular view of the (human) hand; being referred to as the EF-hand (see *Figure 12*).

Equivalent structures have since been found in the intestinal Ca^{2+} binding protein (72), troponin (73) and calmodulin (74) and identified in many proteins on the basis of sequence similarity. Kretsinger (75,76) analysed the essential features of the motif and more recently Gariepy and Hodges (77) have extended this analysis. Two groups

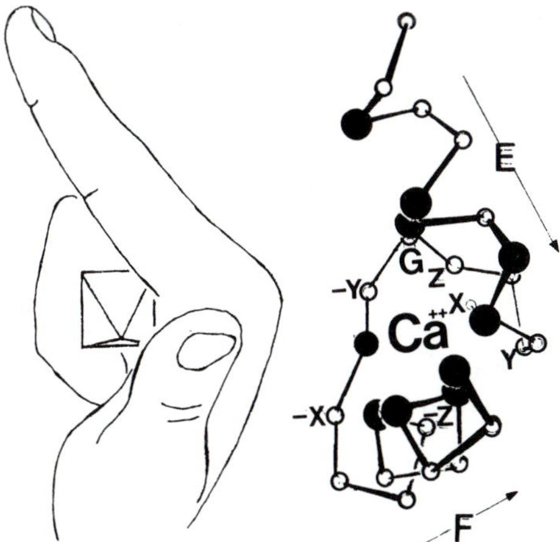

Figure 12. The EF-hand. α-Carbons that contribute hydrophobic side-chains to the core of the protein = ●. The vertices of the octahedron, which represent the oxygen ligands above the Ca^{2+} ion, are indicated by X, Y and Z. The highly conserved Gly = G.

Table 11. Calcium-binding patterns.

(a) EF-hand

Proteins: calmodulin, troponin, intestinal Ca^{2+}-binding protein (ICaBP), parvalbumin, S-100, p11, fibrinogen, SPARC?

Occurrence: repeated motif: two motifs = domain. ICaBP, S-100, p11 = domain; parvalbumin = domain + redundant motif; calmodulin, troponin = two domains. (SPARC predicted to have only one motif).

```
     α1            Ca2+-binding                 α2
αααααααααα            loop              α α αααααααααααα     predominant
--@±-@@p#@D p  B G d G .  @ d  p  E@pp@@.@@.#@              pattern
            X    Y    Z    -Y   -X   -Z
            ss R  E  G D K  ppL  p  K p                      rare variant
                                                             (ICaBP, S-100)
```

Notes: d = a small negative side-chain (mainly aspartic). The positions X,Y,Z and -X,-Y,-Z indicate residues that coordinate the Ca^{2+} ion and the direction of the coordination (see *Figure 12*). The hydrophobic position in the loops is often isoleucine. The conserved G may mutate to K.

(b) Annexins

Proteins: p35 (lipocortin I or calpactin II), p36 (lipocortin II or calpactin I), p32 (endonexin), calelectrin, p10, p68. The repeated domains are preceded by an unconserved stretch (~20 residues) that in some proteins is phosphorylated by a tryosine kinase.

Occurrence: repeated motif: two motifs = domain ? p35 = p36 = endonexin = calelectrin = 2 domains; p10 = 1 domain; p70 = p68 = 4.

```
α1/3   Ca2+-binding   α2/4                                 (predicted structure)
ααααααααα   loop ?   ααααααααα
.Lpp#@...<1>GtD-p.   @@p@@#pR.......@...¥......   pseudo-repeat 1
.Lpp.£.pp-.p G D   @pp@@@s@@  <3-13>              pseudo-repeat 2
```

Notes: position 't' is predominantly threonine but also valine. The variable linker (3−13 residues) is longer in alternate loops.

of sequences were defined by the length of the binding loop the smaller of which (four sequences) contained two extra residues in the loop. Considering the dominant group, a good consensus sequence exists consisting of a loop containing a conserved Gly flanked by alternate residues that bind the Ca^{2+} ion and are well conserved oxygen-containing residues (see *Figure 12* and *Table 11a*).

An interesting possible variant on this motif has recently been identified in the developmental protein SPARC. This Ca^{2+}-sensitive protein contains the sequence pattern with a Lys in place of the very conserved Gly. Modelling of this change showed that such a change created no steric hindrance in the structure (78). In the same protein two of the conserved hydrophobic residues in the helical regions are replaced by Cys at positions suitable for disulphide formation. These changes may be associated with the unusual occurrence of a single motif in this protein while all those described above contain two motifs. It can be imagined that a disulphide bond compensates for loss of hydrophobic packing in the absence of a symmetrically opposed motif.

A further variant of the EF-hand pattern is found in the protein p11. This protein subunit regulates the behaviour of the major substrate (p36) of the tyrosine-specific protein kinases. p11 is homologous to the S-100 protein but in the region equivalent to the longer of the S-100 loops, p11 has a three-residue deletion (79). The interaction of p11 and p36 is interesting as p36 is also Ca^{2+} modulated (see below).

6.2.2 *Annexins*

A new family of Ca^{2+} proteins has recently been identified. These intracellular proteins bind membrane phospholipids in a Ca^{2+}-dependent manner and are generally associated with the cytoskeleton (80,81). The family includes the proteins p35 and p36 both of which are substrates for receptor-associated tyrosine kinases. The family thus appears to fulfil the important role of translating extracellular messages into morphological and physiological cellular responses.

Members of this (functionally defined) family have different molecular weights. The main group lies between 30 and 40 kd with a heavier pair (p68, p70) and one example of a lighter protein (p10). Sequences of the medium weight group (p32, p35, p36 and calelectrin) reveal a 4-fold internal repeat of roughly 70 residues while in the heavier p70 there are eight repeats. The repeats are remotely, but clearly, related and give rise to the consensus sequence shown in *Table 11b*. Like the calmodulin superfamily the two most conserved residues are widely spaced glycines flanked by partially conserved oxygen-containing residues. The pattern of conserved residues in these sequences together with secondary structure predictions has led Taylor and Geisow (4) to propose that this new family may be structurally related to the calmodulin superfamily, differing mainly in the extent of their Ca^{2+} binding loops. Such a model allows most of the conserved features in the consensus sequence to be rationalized.

6.3 **DNA-binding proteins**

A large number of proteins are known to interact directly with DNA. Many of these have had their sequences determined and some have crystallographically determined structures. DNA-binding proteins can be divided as those that bind DNA in some sequence-unspecific structural or enzymic role (e.g. histones or polymerases) and those

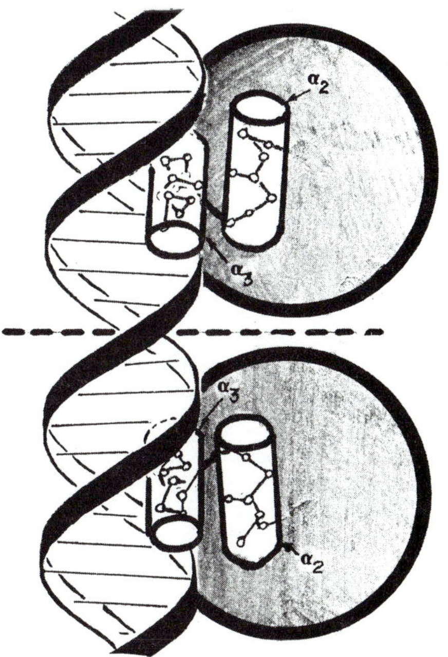

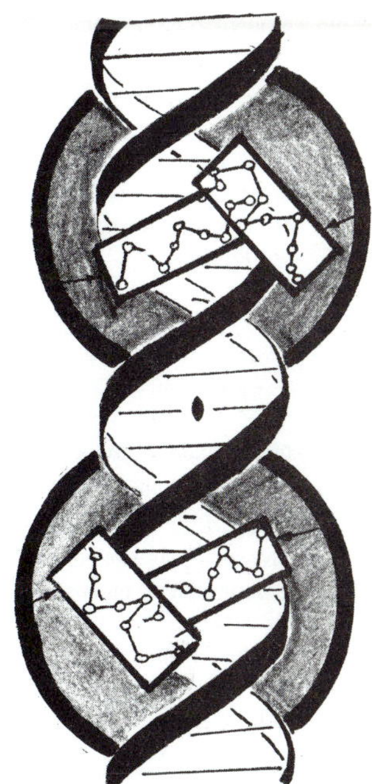

Figure 13. α-Carbons from helices two and three of the proposed Cro–operator complex. Although the rest of the protein structures are quite different, the corresponding helical regions of repressor and CAP are quite similar and may contact the DNA in a similar manner. In each domain α-3 binds in the major groove of the DNA. The molecular 2-fold is indicated as ◆ and ------ about which the two domains (shaded) are related.

that bind in a sequence-dependent manner. In the light of recent sequence alignments (82,83) the DNA pol I polymerases show promise of extending towards a superfamily which is interesting as the group has a member with a known tertiary structure. We will concentrate, however, on the sequence-dependent binders as four of these have known tertiary structures.

6.3.1 *Helix-coil-helix motif*

Proteins that bind specific DNA sequences are often involved in the regulation of gene expression. Those that have known crystallographically determined structures include the cI (lambda-repressor) and Cro (lambda-Cro) repressors of bacteriophage lambda, the CAP of *E.coli* [all reviewed by Palo and Sauer (84)] and the repressor–operator complex of bacteriophage 434 (which is related to lambda). Only the latter structure has been detemined bound to DNA (85).

All these proteins bind as dimers to DNA. The λ-Cro and 434-phage repressors are small proteins consisting largely of α-helical structure. Two of their helices (2 and 3) interact closely with the DNA with α-3 lying along the major groove of DNA (see *Figure 13*). In this orientation the symmetrically related subunit also places its α-3 in an adjacent groove. The λ-repressor, despite its much larger size (236 residues), interacts with DNA in an equivalent manner, as does the CAP protein which is also large (209 residues). The CAP protein has, however, a quite different domain structure to the

Table 12. DNA-binding patterns.

(a) Helix-turn-helix

Proteins: phage (é.g. λ-cro) repressor/operator proteins, catabolite activator protein (CAP), heat-shock regulatory protein, sporulation control proteins, homeo-box proteins, DNA resolvase and inversion proteins
Occurrence: generally in a small all-α domain of variable position in sequence. (Acts as a dimer see *Figure 13*)

```
α2              α3
αααααααα      αααααααα
++p#§..@G@ppppβp+#
```

Notes: many of the less conserved positions interact with DNA and vary because of their different nucleotide sequence specificities but there is a bias towards positive amino acids.

(b) Finger motif

Proteins: transcription factors (e.g. TFIII), developmental control proteins (e.g. Krüppel segmentation gene product), glucocorticoid receptor?, some retroviral proteins.
Occurrence: repeated motif: typically 2–9 repeats. Each repeat thought to bind a Zn^{2+} ion (ligand residues emboldened below).

¥.**C**.<2>.**C**...F.....L..**H**. . .**H**tg...
.**C**. .**C**...F.....L..**H**.....**C**......

Notes: no structure is known. Only conserved identities are indicated.

others. The second and third helices in its DNA binding domain bind the DNA as expected but this domain is preceded by a domain of largely β-structure where the cAMP is bound (see Section 6.1.3).

The close structural similarity of these helix-turn-helix motifs is reflected in a similarity in their sequences (see refs 86–88) and for alignments; summarized in *Table 12*). The most conserved position is a Gly in the tight (three-residue) turn followed by some conserved hydrophobic positions in the flanking helices. Other positions in the helices are variable to accommodate the hydrogen-bonding requirements of different sequence specificity, but in general, positive residues are preferred.

The motif has been identified by (often close) sequence homology in many proteins, including the developmental control 'homeo-box' proteins (89,90).

6.3.2 *Finger motif*

A DNA binding motif has been identified in those sequences that contain a repeated Cys–His pattern. This motif referred to as a 'finger' has a closely grouped pair of cysteines followed after several residues by a closely grouped pair of histidines. Together the Cys and His residues are thought to chelate a zinc ion with the intervening loop extending like a finger to bind DNA (91–93), possibly in an α-helical conformation.

6.4 **Cysteine-rich patterns**

Disulphide bonds are commonly found in extracellular proteins. Their stabilizing effect on the protein fold is especially beneficial to small proteins that might not otherwise

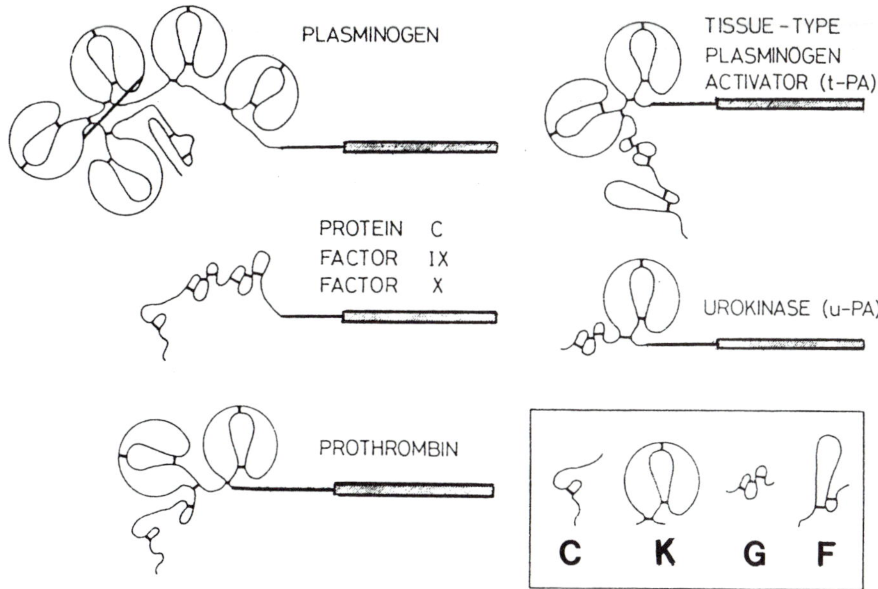

Figure 14. Structures of plasminogen, factor IX, factor X, protein C, prothrombin, urokinase and tissue-type plasminogen activator. The bars represent the protease regions homologous to trypsin. The inset shows the different modules of the non-protease regions. (C): vitamin K-dependent calcium-binding-module, (K): kringle-module, (G): growth factor-module, and (F): finger-module.

generate a sufficiently large hydrophobic core for stability. The sequences of such proteins (examples of which can be found in small toxins and agglutinins) are characteristically cysteine-rich (perhaps 10% Cys: i.e. three disulphides in 60 residues). Although there is no sequence homology across these families, Drenth *et al.* (94) have pointed out that the toxins and agglutinins have a similar tertiary fold.

6.4.1 *Kringle domain*

A relatively Cys-rich domain (three disulphides in 80 residues) is found in proteins associated with the human blood clotting system (e.g factor XII, urokinase, tissue plasminogen activator) (95) and has recently been equated with one of the cysteine rich repeats in fibronectin (96,97). These domains were first identified in prothrombin and generally occur in multiple copies near the N terminus. Several binding functions have been associated with them including binding to fibrin and fibrinogen. It seems likely that the kringles function to direct the activity of the serine protease domain (which they normally precede) to its substrate (see *Figure 14* and *Table 13a*).

Light has recently been shed on the motif by the solution of a fragment of the prothrombin structure (98) in which two of the three disulphides are found tightly packed in the centre of the molecule.

6.4.2 *EGF-like domains*

The sequences of several large extracellular proteins contain cysteine-rich regions that have been postulated to form small, repeated domains. The occurrence of such domains

Table 13. Cysteine-rich domains.

(a) Kringle domain

Proteins: blood-coagulation and fibrolytic factors: prothrombin, plasminogen activator (tPA), plasminogen, urokinase factor (XIIa), fibronectin type II domain and seminal fluid protein.
Occurrence: repeated domain (1 – 11 repeats). In the blood proteins the domains precede a serine proteins domain.

C....G..YRG....T..G..C..W <19> L..NFCRNPD.....PWCYTT <8> C....c

Notes: only conserved identities are shown. The disulphide links above the pattern are buried close together in the protein core and are conserved between the blood proteins and fibronectin. The disulphide link below is less conserved with the final Cys mutable even in the blood proteins.

(b) EGF domain

Proteins: Epidermal growth factor (EGF), transforming growth factor (TGF-α), blood proteins (factors IX and X, tPA, proteins C and S, complement protein C1r), laminin, lin and notch developmental control proteins
Occurrence: single domain in TGF and EGF; multiple repeats in EGF precursor, lin, notch and laminin (laminin repeats only weakly homologous); normally few repeats in blood proteins.

<--------------- large domain ---------------> < --------------- small domain --------------->

ββββ ββββββ βββββ β βββ βββ

.....C <?−?> C <?−?> G..C..<?−?>....C.C..GY.G..C...<?>

Notes: only conserved identifies shown. The two domains fold as simple β-meanders (up, down, up).

is widespread across many functionally distinct families including (again) some blood clotting factors, laminin, epidermal growth factor (EGF) and even its receptor. Alignment of such domains is difficult across families as there is often little conserved except a possible equivalence of cysteines based on characteristic spacings. This is to be expected given the dominant role of the disulphide bridges in maintaining structure, thus allowing the other residues more freedom to mutate.

This superfamily, defined by homology to EGF (99), can be extended from clear homologies [e.g. TGF (100) to more remote relationships such as laminin (101)] in which an extra disulphide must be postulated. In these sequences there is a well conserved 'core' pattern flanked by cysteines of more variable spacing. The determination of the EGF structure using NMR has revealed a largely β-sheet structure (102,103) with two distinct domains. This has shown that the two conserved glycines in the 'core' pattern are conserved for steric reasons (*Table 13b*).

6.5 Immunoglobulin domain

The immunoglobulins (Ig) are a large, structurally well characterized, family of multidomain proteins. Each domain consists of a pair of β-sheets stacked together as a sandwich, the inner faces of which are linked by a conserved disulphide bond. The sequences can all be aligned with varying degrees of confidence revealing that little is conserved over the different domains except for the disulphide cysteines, a tryptophan

Table 14. Immunoglobulin domain.

Proteins: immunoglobulin (Ig) chains, T-cell receptor, MHC proteins cell surface recognition proteins (Thy-1, nCAM),
Occurrence: repeated domains: two types; variable (V) (antigen-binding domain) and constant (C) domain. Ig. two chains; light = V−C, heavy = V−C (N = 3−4). T-cell receptor, two chains; $\alpha = \beta =$ V−C. Other proteins (MHC, β-2 microglobulin, thy-1, nCAM) have less well-defined domain types.

(a) Variable domain

```
              βb          CDR1       βc    βd  βd'      βe        CDR2      βf
        ββββββββββββ                ββββββ             ββββββ           ββββββββββ
  #.#.p#..@.@.C.@ss <??> #.W@.p <??> p.R#$.#.# <??> ..##@.@..@ <??>
               |
               |             βg         CDR3           βh
               |          βββββββ                 ββββββββββββ
               |          s.*#@OC       <??>             ##Gp##.β
               |_______________|
```

Notes: the position '*' is small and hydrophobic. CDR*n* are the antigen binding hypervariable loops.

(b) Constant domain

```
     βb              βc              βd        βe           βf
ββββββββββββ      ββββββ                                ββββββββββ
   @#C.@.s     s@.@.W..p            <??>               βββ....p.#.#.#.@.#
      |
      |            βg                βh
      |          βββββββ           ββββββ
      |          O.C.@.pp.  <??>  #@.+##....(C)
      |____________|
```

Notes: in the LC and CH1 domains (see *Figure 15a*) the final cysteine (C) is conserved to form the inter-chain disulphide bond. In Thy-1 and the β-2 microglobulin the conserved W mutates to a large hydrophobic residue. It is suspected that the very conserved disulphide might also be mutable in some poly Ig proteins (see *Figure 15b*). Regions indicated <??> may vary in length by several residues.

and scattered hydrophobic residues (10) (see *Table 14*).

The immunoglobulin domain can also be clearly recognized in sequences of proteins associated with the cellular immune response (104) including T-cell receptors and the proteins of the major histocompatibility complex (MHC). The T-cell receptor appears to be composed of typical Ig domains while the MHC cell surface antigen-presenting proteins (e.g. HLA) are only partially Ig like. In the MHC proteins found on all cells (class I) a membrane-bound Ig domain is combined with a larger domain and a small Ig subunit (β-2 microglobulin), while in the equivalent proteins predominantly associated with T-cells (class II) there are two membrane-bound chains each with an Ig domain next to the membrane (see *Figure 15*).

A further occurrence of a remotely related Ig domain is found in some other cell surface proteins, including Thy-1, a single membrane-bound glycosylated Ig-like domain, and the neural cell adhesion molecule (nCAM) which is composed of repeated Ig domains like the poly Ig T-cell receptor.

7. SPECIFIC FAMILY PATTERNS

The patterns discussed above can be considered as general structural units that can be incorporated as a domain, or part of a domain, into a larger structure. This modular structure is typified by trypsin plasminogen activator which is a linear assembly of a fibronectin-like 'finger' module, an EGF-like domain and two kringle domains followed by a serine protease (see *Figure 14*).

Although the borderline between general motifs and patterns specific to a single family is, to a large extent, arbitrary, most of the remaining patterns and motifs that could be discussed are peculiar to a single family. As it would be an immense task to characterize consensus sequences for all known protein families, we will instead consider just one example to illustrate how patterns specific to a single family can be derived and used predictively on sequences of unknown structure.

The example we will consider is the family of aspartyl proteases. This is one of the major families of proteases (others include the serine, thiol and metallo proteases) and includes three members with crystallographically determined structures. The aspartyl proteases are particularly interesting as they contain a remotely related duplication in their sequences that can be seen in the structures to correspond to an approximate 2-fold symmetry. The sequence duplication is easily identified by the absolutely conserved residues DTG which occur some 30 residues from the N terminus of each domain.

7.1 Remote sequence alignment

An alignment of all members of the acid protease family is possible over almost the entire length of the sequence. However, except for the DTG regions, the alignment

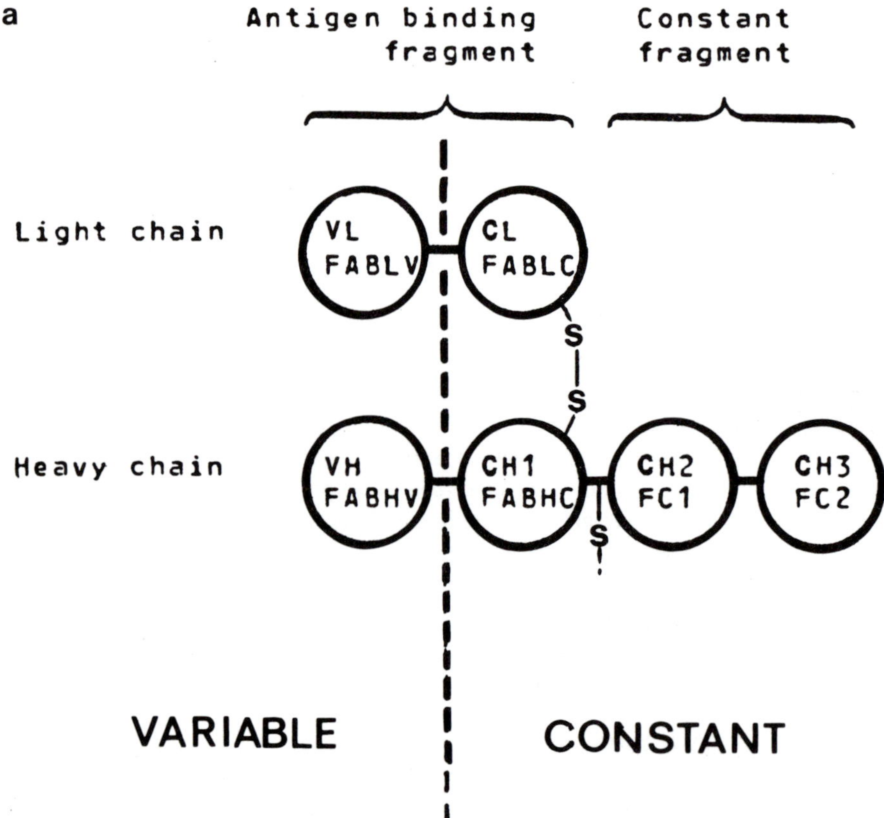

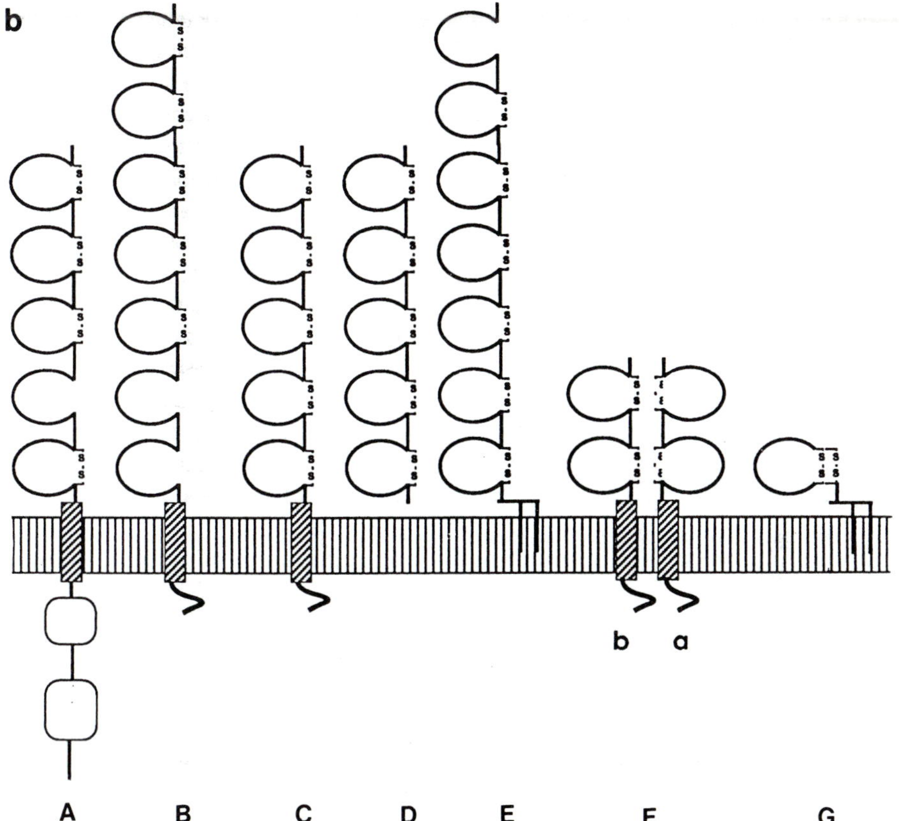

Figure 15. Members of the Ig superfamily, including NCAM (B) (with a short cytoplasmic tail), MAG (C) domain organization similar to the poly Ig receptor, α1-B glycoprotein (D), CEA (E) T-cell receptor (F) and thy-1 (G). The Ig homology units are represented as disulphide-bonded loops.

of each internally repeated domain with the other can only be accomplished with reference to the topological equivalence defined by the X-ray structures. Such a remote alignment poses special problems for the definition of a consensus sequence. Ideally a consensus pattern should be able to reproduce the alignment from which it was derived when it is matched against each of those sequences individually. A pattern with this property will be referred to as 'consistent'.

Unfortunately, the patterns characterizing an alignment of remotely related sequences are very often inconsistent as there will generally be more than one possible location for each pattern. To encourage consistency, constraints can be applied to the matching: clearly the sequential order of the patterns should be conserved and also, if necessary, the size of gap allowed between them. This approach of constrained pattern matching was used by Taylor in testing consensus pattern matching methods on the immunoglobulin sequences (10).

With the aspartyl protease sequences, the conserved DTG sequence (along with flanking hydrophobic residues) provides a good 'marker' in the sequence relative to which the remainder of the domain should be located. The second best conserved pattern is a OOG sequence (O = hydrophobic) that is expected some 70 residues C-terminal to the DTG. In some of the sequences the OOG pattern is not unique but when limits are placed on its distance from the DTG pattern a unique (and correct) position can be found in all the sequences. Thus, (as with the nucleotide binding patterns in the kinases) two sequentially remote conserved patterns have defined anchor points in a sequence alignment that can then be extended by conventional alignment methods, additional consensus patterns or simply by the equivalence of predicted secondary structures. These patterns are surprisingly specific for aspartyl proteases and when matched over the protein identification sequence (PIR) database, identify no proteins that are not known (or thought to be) aspartyl proteases. Interestingly, regions in the polyproteins encoded by the retroviruses are also recognized by this pattern. These too have a conserved DTG sequence followed some 50 residues later by a conserved OOGRD sequence (105).

7.2 **Building a molecular model**

In the tertiary structure of the aspartyl proteases these two patterns lie close to the active site at the ends of adjacent β-strands (again, equivalent to the kinase patterns) and thus define the core of the structure. This feature was used to propose a structure for one of the remotely related retroviral proteases (from HIV-1) as it allowed a core to be defined with reasonable certainty relative to which the remaining sequence was located by less certain secondary structure prediction methods including the prediction of β-turns (105) (see *Figure 16*).

This example illustrates how different parts of a model-built structure must be viewed with differing degrees of confidence. For example, the absolute conservation of the residues in the active site imply that the geometry of the site has been conserved throughout evolution, thus the model in this region implies a structural prediction at the atomic level. The secondary structure of the core supporting the active site residues is probably conserved with identical topology, as in all domains of known structure in the family these features are conserved. However, most of the residues have undergone conservative mutation and the loop lengths away from the active site have varied in length. Beyond this, on the edges of the β-sheet, the secondary structure predictions still indicate the presence of β-strands but there is little or no sequence correspondence with a protein of known structure so here, even the chain topology is uncertain.

8. CONCLUSIONS

In this chapter we have attempted to survey in proteins the known patterns of structural interest. However, as the number of known protein sequences increases rapidly, new families are continually identified and the old extended and with new crystal structures, patterns that were previously known only from the sequence, find a structural context. Because of this continual expansion in our knowledge, any survey rapidly becomes dated. Thus, rather than hope to be comprehensive, we have intended this chapter to present a guide to the basic patterns from which investigators can develop their own variants either directly or by application of the methodologies described.

Ab initio structure prediction has advanced very slowly since the first attempts in the 1960s. Recently the approach has been recognition of known folds and sequence patterns, which can be used as the starting point from which a reliable prediction can

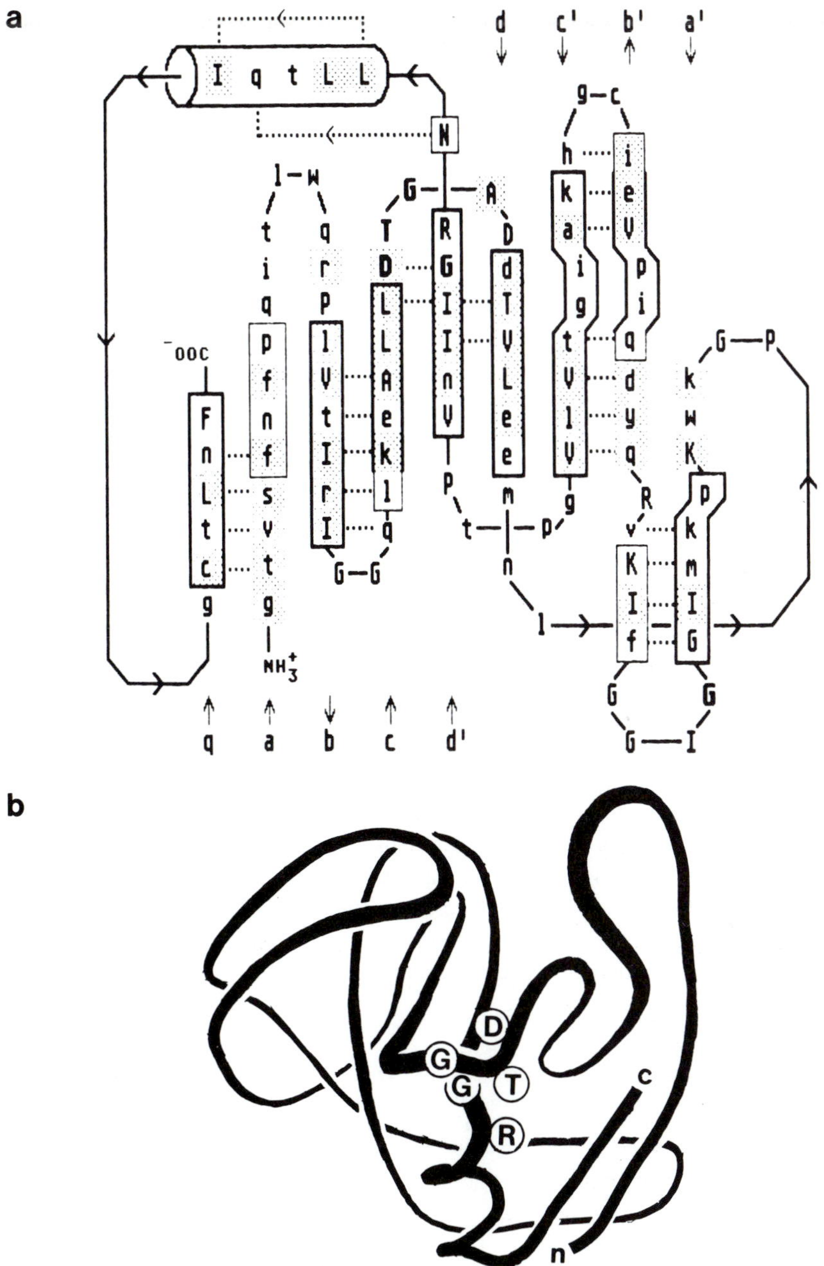

Figure 16. (**a**) Schematic diagram of the predicted HIV-1 *pol* protease fold. The amino acid sequence is given as single letter codes, with lower case indicating residues specific to HIV-1, upper case indicating residues conserved in retroviral proteases, and bold upper case indicating residues also conserved in aspartic proteases. Boxes are predicted β-strands, with strong predictions indicated by heavy walls. A predicted α-helix is shown as a cylinder. Residues with a low solvent accessibility have a speckled background. (**b**) Predicted tertiary structure of HIV-1 *pol* protease. Diagram of the *pol* protease fold, showing the positions of the residues most highly conserved in the viral sequences.

be made. We have currently more than 14 000 sequences and only 200–300 structures. A critical question is, is there a limited number of folds, and if so, how many do we know? It is likely that there will be a limited set of supersecondary folds, which it may be possible to identify. These patterns will be stored in databases and become accessible to the scientific community with standard search programmes. At Birkbeck and Leeds Universities we have already established the basic structural and sequence databases using Oracle and we are working towards encoding the sequence and structural features. This work is funded by the Protein Engineering Club and the database is now available for general use.

It can be imagined that, given the combined sequence and structural databases, an automaton could be devised that would continually align sequences and identify patterns. One of the impediments to executing such a project is finding an adequate description for the patterns that have been identified. As can be seen from the tables in this chapter, such descriptions are often vague (allowing exceptions in some places and not others) and inconsistent between different patterns. However, proteins are so complex that no simple method of pattern definition can be sufficiently flexible to capture the subtleties of amino acid requirement and allow variation at each position. Furthermore the description must also capture the interaction between positions within and between patterns and ultimately this implies that the protein must be viewed in its entirety, including its biological context.

For the patterns described here, it is our current aim to develop templates that approach the two criteria described in the preceding section: they should be both consistent and unique. In other words they should relocate their correct position on each sequence that contributed to their definition and should be specific only to members of that group. To approach this requires that the method of template specification has sufficient flexibility along the lines described above and, using the method of Taylor (10) we have refined several patterns towards these ideals. Luckily, such a complex system is not necessary for identifying most of the patterns described above, nevertheless we are continuing to work towards a more powerful method of finding patterns in proteins as this, we believe, holds the key to the automatic interpretation of protein sequence data, including the prediction of tertiary structure.

9. REFERENCES

1. Chou,P.Y. and Fasman,G.D. (1978) *Adv. Enzymol.*, **47**, 45.
2. Garnier,J., Osguthorpe,J.D. and Robson,B. (1978) *J. Mol. Biol.*, **120**, 97.
3. Taylor,W.R. (1987) In *Nucleic Acid and Protein Sequence Analysis: A Practical Approach.* Bishop,M.J. and Rawlings,C.J. (eds). IRL Press, Oxford, p. 285.
4. Taylor,W.R. and Geisow,M.J. (1987) *Protein Engin.*, **1**, 183.
5. Bishop,M.J., Ginsberg,M., Rawlings,C.J. and Wakeford,R. (1987) In *Nucleic Acid and Protein Sequence Analysis: A Practical Approach.* Bishop,M.J. and Rawlings,C.J. (eds). IRL Press, Oxford, p. 19.
6. Barker,W.C., Hunt,L.T., Orcutt,B.C., George,D.G., Yeh,L.S., Chen,H.R., Blomquist,M.C., Johnson,G.C., Seibel-Ross,E.I., Hong,M.K. and Ledley,R.S. (1984) *Protein Identification Resource.* National Biomedical Research Federation. Washington, DC, USA.
7. Orcutt,B.C., George,D.G., Fredrickson,J.A. and Dayoff,M.O. (1982) *Nucleic Acids Res.*, **10**, 157.
8. Wilbur,W.J. and Lipman,D.J. (1983) *Proc. Natl. Acad. Sci. USA*, **80**, 726.
9. White,S. (1989) *Protein Eng.*, in press.
10. Taylor,W.R. (1986) *J. Mol. Biol.*, **188**, 233.
11. Taylor,W.R. (1987) *CABIOS*, **3**, 81.
12. Lim,V. (1974) *J. Mol. Biol.*, **88**, 873.

13. Eliopoulos,E. and Geddes,A.J. *Protein Structure Prediction Suite*. Department of Biophysics, University of Leeds, Leeds LS2 9JT, UK.
14. Lewis,P.N., Momany,F.A. and Scheraga,H.A. (1971) *Proc. Natl. Acad. Sci. USA*, **68**, 2293.
15. Venkatachalan,C.M. (1968) *Biopolymers*, **6**, 1425.
16. Richardson,J.S. (1983) *Adv. Protein Chem.*, **34**, 167.
17. Wilmot,C.M. and Thornton,J.M. (1988) *J. Mol. Biol.*, **203**, 221.
18. Levine,J.M., Robson,B. and Garnier,J. (1986) *FEBS Lett.*, **205**, 303.
19. Nishikawa,K. and Ooi,T. (1986) *Biochim. Biophys. Acta*, **871**, 45.
20. Zvelebil,M.J., Barton,G.J., Taylor,W.R. and Sternberg,M.J.E. (1987) *J. Mol. Biol.*, **195**, 957.
21. Kabsch,W. and Sander,C. (1983) *FEBS Lett.*, **155**, 179.
22. Nishikawa,K. (1983) *Biochim. Biophys. Acta*, **748**, 285.
23. Kabsch,W. and Sander,C. (1984) *Proc. Natl. Acad. Sci. USA*, **81**, 1075.
24. Garratt,R.C., Taylor,W.R. and Thornton,J.M. (1985) *FEBS Lett.*, **188**, 59.
25. Levitt,M. and Chothia,C. (1976) *Nature*, **216**, 552.
26. Taylor,W.R. and Thornton,J.M. (1984) *J. Mol. Biol.*, **173**, 487.
27. Nishikawa,K. and Ooi,T. (1982) *J. Biochem.*, **91**, 1821.
28. Rose,G.D., Geselowitz,A.R., Lesser,G.J., Lee,R.H. and Zehfus,M.H. (1985) *Science*, **229**, 834.
29. Wolfenden,R., Anderson,L., Cullis,P.M. and Southgate,C.C. (1981) *Biochemistry*, **20**, 849.
30. Fauchere,J.L. and Pliska,V. (1983) *Eur. J. Med. Chem.*, **10**, 369.
31. Miller,S.M., Janin,J., Lesk,A.M. and Chothia,C. (1987) *J. Mol. Biol.*, **157**, 105.
32. Kyte,J. and Doolittle,R.F. (1982) *J. Mol. Biol.*, **157**, 105.
33. Eisenberg,D. (1984) *Annu. Rev. Biochem.*, **53**, 595.
34. Janin,J. (1979) *Nature*, **277**, 491.
35. Kuntz,I.D. (1972) *J. Mol. Chem. Soc.*, **94**, 4009.
36. Rose,G.D. (1978) *Nature*, **272**, 586.
37. Hopp,T.P. and Woods,K.R. (1981) *Proc. Natl. Acad. Sci. USA*, **78**, 3824.
38. Karplus,P.A. and Schultz,G.E. (1985) *Naturwissenschaften*, **72**, 5212.
39. Thornton,J.M., Edwards,M.S. and Barlow,D.J. (1985) In *Conference Proceedings for Computer Aided Molecular Design*. Basel.
40. Novotný,J., Handschumacher,M. and Haber,E. (1986) *J. Mol. Biol.*, **189**, 715.
41. Thornton,J.M., Edwards,M.S., Taylor,W.R. and Barlow,D.J. (1986) *EMBO J.*, **5**, 409.
42. Argos,P.A., Rao,J.M. and Hargrave,P.A. (1982) *Eur. J. Biochem.*, **128**, 565.
43. Schiffer,M. and Edmundson,A.B. (1968) *Biophys. J.*, **8**, 29.
44. Finer-Moore,J. and Stroud,R.M. (1984) *Proc. Natl. Acad. Sci. USA*, **81**, 155.
45. Cohen,F.E., Sternberg,M.J.E. and Taylor,W.R. (1982) *J. Mol. Biol.*, **156**, 821.
46. Chothia,C. and Janin,J. (1980) *J. Mol. Biol.*, **143**, 215.
47. Richmond,T.J. and Richards,F.M. (1978) *J. Mol. Biol.*, **119**, 537.
48. Sibanda,B.L. and Thornton,J.M. (1985) *Nature*, **316**, 170.
49. Edwards,M.S., Sternberg,M.J.E. and Thornton,J.M. (1987) *Protein Eng.*, **1**, 173–181.
50. Efimov,A. (1986) *Mol. Biol.*, **20**, 329.
51. Thornton,J.M., Sibanda,B.L., Edwards,M.S. and Barlow,D.J. (1988) *Bioessays*, **8**, 63.
52. Hodgman,T.C. (1986) *CABIOS*, **2**, 181.
53. Rossman,M.G., Liljas,A., Bränden,C.-I. and Banaszak,L.J. (1975) In *The Enzymes*, Boyer,P. (ed.). Academic Press, NY, Vol. 11, p. 61.
54. Bränden,C.-I. (1980) *Q. Rev.Biochem.*, **13**, 317.
55. Wierenga,R.K. and Hol,W.G.J. (1983) *Nature*, **302**, 842.
56. Hol,W.G.J., Van Duijnen,P.Th. and Berendsen,H.J.C. (1987) *Nature*, **273**, 443.
57. Sternberg,M.J.E. and Taylor,W.R. (1984) *FEBS Lett.*, **175**, 387.
58. Moller,W. and Amons,R. (1985) *FEBS Lett.*, **186**, 1.
59. Wierenga,R.K., Terpstra,P. and Hol,W.G.J. (1986) *J. Mol. Biol.*, **187**, 101.
60. Wierenga,R.K., De Maeyer,M.C.H. and Hol,W.G.J. (1985) *Biochemistry*, **24**, 1346.
61. Walker,J.E., Saraste,M., Runswick,W.J. and Gay,N.J. (1982) *EMBO J.*, **1**, 945.
62. Schultz,G.E., Elzinga,E., Marx,F. and Schirmer,R.H. (1974) *Nature*, **750**, 120.
63. Fry,D.C., Kuby,S.A. and Mildvan,A.S. (1986) *Biochemistry*, **24**, 4680.
64. Jurnak,F. (1985) *Science*, **320**, 32.
65. Duncan,M., Parsonage,D. and Senior,A.E. (1986) *FEBS Lett.*,**208**, 1.
66. Taylor,W.R. and Green,N.M. (1989) *Eur. J. Biochem.*, in press.
67. Lochrie,M.A., Hurley,J.B. and Simon,M.I. (1985) *Science*, **228**, 96.
68. Masters,S.B., Stroud,R.M. and Bourne,R.H. (1986) *Protein Eng.*, **1**, 47.
69. Sternlicth,H., Yaffe,M.B. and George,W.F. (1987) *FEBS Lett.*, **214**, 226.
70. Weber,I.T., Steitz,T.A., Bubis,J. and Taylor,S.S. (1987) *Biochemistry*, **26**, 343.

71. Moews,P.C. and Kretsinger,R.H. (1975) *J. Mol. Biol.*, **91**, 201.
72. Szebenyi,D.M.E. and Moffat,K. (1986) *J. Biol. Chem.*, **261**, 8761.
73. Hertzberg,O. and James,M.N.G. (1985) *Nature*, **313**, 653.
74. Babu,Y.S., Sack,J.S., Greenbough,T.S., Bugg,C.E., Means,A.R. and Cook,W.J. (1985) *Nature*, **315**, 37.
75. Tufty,R.M. and Kretsinger,R.H. (1975) *Science*, **187**, 167.
76. Kretsinger,R.H. (1980) *Crit. Rev. Biochem.*, **8**, 119.
77. Gariepy,J. and Hodges,R.S. (1983) *FEBS Lett.*, **160**, 1.
78. Engel,J., Taylor,W.R., Paulsson,M., Sage,H. and Hogan,B. (1987) *Biochemistry*, **26**, 6958.
79. Gerke,V. and Weber,K. (1985) *EMBO J.*, **4**, 2917.
80. Owens,R.J. and Crumpton,M.J. 1984) *Bioessays*, **1**, 61.
81. Geisow,M.J. and Walker,J.H. (1986) *Trends Biochem. Sci.*, **11**, 420.
82. Argos,P., Tucker,A.D. and Philipson,L. (1986) *Virology*,**149**, 208.
83. Ollis,D.L., Kline,C. and Steitz,T.A. (1985) *Nature*, **313**, 818.
84. Pabo,C.O. and Sauer,R.T. (1984) *Annu. Rev. Biochem.*, **53**, 293.
85. Anderson,J.E., Patashne,M. and Harrison,S.C. (1987) *Nature*, **326**, 846.
86. Anderson,W.F., Takeda,Y., Ohlendorf,D.H. and Mathews,B.W. (1982) *J. Mol. Biol.*, **159**, 745.
87. Dodd,I.B. and Egan,J.B. (1987) *J. Mol. Biol.*, **194**, 557.
88. Argos,P. (1981) *J. Theor. Biol.*, **93**, 609.
89. Shepherd,J.C.W., McGinnis,W., Carrasco,A.E., De Robertis,E.M. and Gehring,W.J. (1984) *Nature*, **310**, 70.
90. Laughon,A. and Matthews,P.S. (1984) *Nature*, **310**, 25.
91. Chowdhury,K., Deutsch,U. and Gruss,P. (1987) *Cell*, **48**, 771.
92. Rosenberg,U.B., Schröder,C., Preiss,A., Kienlin,A., Côté,I.R., Riede,I. and Jäckle,H. (1986) *Nature*, **319**, 336.
93. Miller,J., McLachlan,A.D. and Klug,A. (1985) *EMBO J.*, **4**, 1609.
94. Drenth,J., Low,B.W., Richardson,J.S. and Wright,C.S. (1980) *J. Biol. Chem.*, **255**, 2652.
95. Magnusson,S., Peteren,T.E., Sottrup-Jensen,L. and Blaeys,H. (1975) In *Proteases and Biological Control.* Reich, Rifkin and Shaw (eds). Cold Spring Harbor Laboratory Press, Cold Spring Harbor, NY, p. 123.
96. Skortstengaard,K., Jensen,M.S., Sahl,P., Petersen,T.E. and Magnusson,S. (1986) *Eur. J. Biochem.*, **161**, 441.
97. Holland,S.K., Harlos,K. and Blake,C.C.F. (1987) *EMBO J.*, **6**, 1875.
98. Park,C.H. and Tulinsky,A. (1986) *Biochemistry*, **25**, 3977.
99. Savage,C.R., Hash,J.H. and Cohen,S. (1973) *J. Biol. Chem.*, **248**, 7669.
100. Marquardt,H., Hunkapilleir,M.W., Hood,L.E. and Todaro,G.J. (1984) *Science*, **223**, 1079.
101. Sasaki,M., Kato,S., Kohno,K., Martin,G.R. and Yamada,Y. (1987) *Proc. Natl. Acad. Sci. USA*, **84**, 935.
102. Cooke *et al.* (1987) *Nature*, **327**, 339.
103. Montelione,G.T., Wütrich,K., Nice,E.C., Burgess,A.W. and Scheraga,H.A. (1986) *Proc. Natl. Acad. Sci. USA*, **83**, 8594.
104. Kabat,E.A., Wu,T.T., Reid-Miller,M., Perry,H.M. and Gottesman,K.S. (1987) *Sequences of Proteins of Immunological Interest.* National Institues of Health, Bethesda, MD, USA.
105. Pearl,L.H. and Taylor,W.R. (1987) *Nature*, **329**, 351.
106. Akrigg,D., Bleasby,A.J., Dix,N.I.M., Findlay,J.B.C., North,A.C.T., Parry-Smith,D., Wootton,J.C., Blundell,T.L.B., Gardner,S.P., Hayes,F., Islam,S., Sternberg,M.J.E., Thornton,J.M., Tickle,I.J. and Murray-Rust,P. (1988) *Nature*, **335**, 745.

Protein modifying agents

Amino acid	*Reagent*	*Conditions*	*Reference*[a]
Arginine	2,3-Butanedione	pH 7–8, 100-fold molar excess, 25°C, 1 h Can react with amino groups	1
	Camphorquinone-10-sulphonic acid	pH 9, 37°C, up to 24 h, reversible	2
	1,2-cyclohexanedione	pH 9, 35°C, 2 h	3
	p-OH and *p*-nitro phenylglyoxals	pH 7–9, 50 μm, 20–30°C, 30 min	4
Aparagine	Bis-(1,1-trifluoroacetoxy) iodobenzene	pH 2–3, 50 mM, 60°C, 4 h Also reacts with several other amino acids	5
Aspartic acid	Carbodiimides	pH 5, up to 100 mM, 20–40°C, 1–2 h	6
Cysteine	Acetic anhydride	As for Lys	
	Bromo-, chloro- and iodoacetic acid/acetamide	pH 8–9, up to 10 mM, 30–60 min, 4–30°C	7,8
	p-Chloromercuribenzoate or benzene sulphonic acid	pH 5–7, up to 1mM, 20–30°C, up to 1 h	9,10
	Dansyl and dabsyl chlorides	As for Lys	11
	5,5′-Dithiobis-(2-nitrobenzoic acid) (Ellmans Reagent)	pH 7.3–8.0, μm, 25°C, up to 1 h	12
	N-(Iodophenyl) trifluoroacetamide	pH 8, up to 50-fold excess, 20–50°C, up to 1 h	13
	Methyl iodide	pH 8, μM, 20°C, 15 min Also reacts with Met	41
	N-Ethyl and phenyl maleimides	pH 7–8, up to 1 mM, 20–37°C for up to 1 h	14
	Ammonium 4-chloro-7-sulphobenzofurazan (SBF-chloride)	pH 8, mM, 30°C, for up to 3 h	15
	3,5-Diiodo-4-diazobenzene sulphonic acid	pH 7.4, 1 mM, 4–25°C, 5–30 min	16
	4-Vinylpyridine	See Chapters 1 and 4	
Glutamic acid	Carbodiimides	As for Asp	6

Amino acid	*Reagent*	*Conditions*	*Reference*[a]
Glutamine	Bis-(1,1-trifluoroacetoxy) iodobenzene	As for Asn	5
	Transglutaminase	pH 7.4, 25–37°C, for up to 18 h	17
Histidine	Dansyl and dabsyl chlorides	As for Lys	11
	α-Haloacids	pH 5.5–6.5, mM, 20°C, up to 1 h	18
	Ethoxyformic anhydride (also called diethyl pyrocarbonate)	pH 6, up to 1 mM, 25°C, 15 min See also Trp	35
	3,5-Diiodo-4-diazobenzene sulphonic acid	As for Cys	16
Lysine	Choline, ethyl, isethionyl methyl acetimidates	pH 8.0, 1–10 mM, 20–30°C, up to 2 h	20
	Acetic, citraconic, maleic and succinic anhydrides	pH 7.0–7.5, mM, 4–25°C, 10–60 min	10,19 p. 144
	Methyl-3,5-diido and methyl *p*-hydroxybenzimates (Woods reagent)	pH 8–9, μM, 20–40°C, 1 h	21
	Dansyl and dabsyl chloride	As for Cys	11
	Ethyl thiotrifluoroacetate	pH 10, μM, 25°C, 1–2 h	22,23
	1-Fluoro-2,4-dinitrobenzene (Sanger's Reagent)	pH 8–9, up to 10 mM, 20–30°C for 1–3 h. Also reacts with Cys and Tyr	24
	N-Formyl methionyl sulphone methyl phosphate ([^{35}S]FMMP)	pH 9.5–10.0, 50 μm, 20–30°C, up to 30 min	25
	2-Iminothiolane (Traut's Reagent)	Donates free thiol pH 8.0, mM, 4–25°C, 1 h	26
	Isothiocyanates	pH >7.0, up to 1 mM, 4–30°C, 1 h	27,28,39
	O-Phthalaldehyde	pH 9.0–10.5, up to 10 mM, 10–50°C, up to 10 min 2-Mercaptoethanol required for fluorophore	29
	Pyridoxal phosphate	pH 7.5, up to 1 mM, 4–25°C, up to 30 min.	30
	Succinimidyl esters	pH 6–9, μM, 0–30°C, 10–30 min	31 Pierce Amersham
	Succinimidyl propionates (Bolton and Hunter Reagents)	pH 8–9, up to 100-fold molar excess (μM), 0–30°C, 10–30 min	32
	2,4,6-Trinitrobenzene sulphonate	pH 8.0–9.0, 1 mM, 20–40°C, up to 24 h	33

Amino acid	Reagent	Conditions	Reference[a]
Threonine/Serine	Acetic and succinic anhydrides	As Lys but inefficient	34
Tryptophan	Ethoxyformic (acid) anhydride	As for His, will also react with Ser and Tyr	35
	2-nitrophenylsulphenyl chløride	pH 3.0, 4-fold molar excess, 25°C, up to 6 h	36
Tyrosine	Acetic anhydride	As for Lys	
	Dansyl and dabsyl chloride	As for Lys	
	Iodination	Numerous methods	19
	3,5-diido-4-diazobenzene sulphonic acid	As for Lys	
	p-Nitrobenzene sulphonyl chloride	pH 7–8.0, 1 mM, 25°C, 30 min	37
Other useful reagents			
Cys, Met, Tyr, Trp, His, Lys	Aryl azides	Activated by irradiation pH 7–9, μM, –100 to 25°C, 1–60 min	19,38
Non-selective	Diazirines	Activated by irradiation pH 6–8, μM, 0–37°C for up to 20 min	19,38
N-terminus and ϵ-NH_2	Fluorescein isothiocyanate	As for PITC, fluorescent	39,40
All	Cross-linking agents	Utilizing specificities and conditions outlined above	19

[a]REFERENCES

1. Yankeelov,J.A., Jr (1972) *Methods in Enzymology.* Hirs,C.H.W. and Timasheff,S.N. (eds), Academic Press Inc., New York, Vol. 25, p. 566.
2. Pande,C.S., Pelzig,M. and Glass,J.D. (1980) *Proc. Natl. Acad. Sci. USA,* **77**, 896.
3. Patthy,L. and Smith,E.L. (1975) *J. Biol. Chem.,* **250**, 557.
4. Yamasaki,R.B., Shimer,D.A. and Feeney,R.E. (1981) *Anal. Biochem.,* **111**, 220.
5. Soby,L.M. and Johnson,P. (1981) *Anal. Biochem.,* **113**, 149.
6. Carraway,K.L. and Koshland,D.E. (1972) *Methods in Enzymology.* Hirs,C.H.W. and Timasheff,S.N. (eds), Academic Press Inc., New York, Vol 25, p.616.
7. Crestfield,A.M., Moore,S. and Stein,W.H. (1963) *J. Biol. Chem.,* **238**, 622.
8. Gurd,F. (1972) *Methods in Enzymology.* Hirs,C.H.W. and Timasheff,S.N. (eds), Academic Press Inc., New York, Vol. 25, p. 424.
9. Benesch,R. and Benesch,R.E. (1962) *Methods Biochem. Anal.,* **10**, 43.
10. Glazer,A.N., Delange,R.J. and Sigman,D.S. (1976) In *Biochemistry and Molecular Biology.* Work,T.S. and Work,E. (eds), Elsevier, Amsterdam, Vol. 4.
11. Schmidt-Ullrich,R., Knüfermann,H. and Wallach,D.F.H. (1973) *Biochim. Biophys. Acta,* **307**, 353.
12. Riddles,P.W., Blakeley,R.L. and Zerner,B. (197 *Anal. Biochem.,* **94**, 75.
13. Schwartz,W.E., Smith,P.K. and Royer,G.P. (1980) *Anal. Biochem.,* **106**, 43.
14. Riordan,J.F. and Vallee,B.L. (1972) *Methods in Enzymology.* Hirs,C.H.W. and Timasheff,S.N. (eds), Academic Press Inc., New York, Vol. 25, p. 449.
15. Andrews,J.L., Ternal,B. and Whitehouse,M.W. (1982) *Arch. Biochem. Biophys.,* **214**, 386.
16. Barclay,P.L. and Findlay,J.B.C. (1984) *Biochem. J.,* **220**, 75.

17. Pober,J.S., Iwanij,V., Reich,E. and Stryer,L. (1978) *Biochemistry,* **17**, 2163.
18. Heinrikson,R.L., Stein,W.H., Crestfield,A.M. and Moore,S. (1965) *J. Biol. Chem.,* **240**, 2921.
19. Findlay,J.B.C. (1987). In *Biological Membranes: A Practical Approach.* Findlay,J.B.C. and Evans,W.H. (eds), IRL Press, Oxford, p. 179.
20. Nemes,P.P., Miljamich,G.P., White,D.L. and Dratz,E.A. (1980) *Biochemistry,* **19**, 2067.
21. Bright,G.R. and Spooner,B.S. (1983) *Anal. Biochem.,* **131**, 301.
22. Goldberger,B.F. (1967) *Methods in Enzymology.* Hirs,C.H.W. (ed.), Academic Press Inc., New York, Vol. 11, p. 317.
23. Levy,D. and Paselk,R.A. (1973) *Biochim. Biophys. Acta,* **310**, 398.
24. Birnbaumer,M.E., Schrader,W.T. and O'Malley,B.W. (1979) *Biochem. J.,* **181**, 201.
25. Bretscher,M. (1971) *J. Mol. Biol.,* **58**, 775.
26. Lambert,J.M., Baileau,G., Cover,J.A. and Traut,R.R (1983) *Biochemistry,* **22**, 3913.
27. Maddy,A.H. (1964) *Biochim. Biophys. Acta,* **88**, 390.
28. Cabantchik,Z.I. and Rothstein,A. (1974) *J. Membrane Biol.,* **15**, 227.
29. Benson,J.R. and Hare,P.E. (1975) *Proc. Natl. Acad. Sci. USA,* **72**, 619.
30. Rifkin,D.B., Combans,R.W. and Reich,R. (1972) *J. Biol. Chem.,* **247**, 6432.
31. Lomant,A.J. and Fairbanks,G. (1976) *J. Mol. Biol.,* **104**, 243.
32. Bolton,A.E. and Hunter,W.M. (1973) *Biochem. J.,* **133**, 529.
33. Plapp,B.V., Moore,S. and Stein,W.H. (1971) *J. Biol. Chem.,* **246**, 939.
34. Allen,G. and Harris,J.I. (1976) *Eur. J. Biochem.,* **62**, 601.
35. Tsurushiin,S., Hirawatsu,A., Inamatsu,M. and Yasunolsu,K.T. (1975) *Biochim. Biophys. Acta,* **400**, 451.
36. Fontana,A. and Scoffone,E. (1972) *Methods in Enzymology.* Hirs,C.H.W. and Timasheff,S.N. (eds), Academic Press Inc., New York, Vol. 25, p. 482.
37. Liao,T.-H., Ting,R.S. and Yeung,J.E. (1982) *J. Biol. Chem.,* **257**, 5637.
38. Bayley,H. (1983) In *Biochemistry and Molecular Biology.* Work,T.S. and Burdon,R.H. (eds), Elsevier, Amsterdam, Vol. 12.
39. Muramoto,K., Kawauchi,H. and Tuzimura,K. (1978) *Agric. Biol. Chem.,* **42**, 1559.
40. Muramoto,K., Kamiya,H. and Kawauchi,H. (1984) *Anal. Biochem.,* **141**, 446.
41. Rochat,C., Rochat,H. and Edman,P. (1970) *Anal. Biochem.,* **37**, 259.

INDEX